STP 1081

Geosynthetic Testing for Waste Containment Applications

Robert M. Koerner, editor

ASTM
1916 Race Street
Philadelphia, PA 19103

Library of Congress Cataloging-in-Publication Data

Geosynthetic testing for waste containment applications / Robert M.
    Koerner, editor
        STP 1081
        Papers presented at the Symposium on Geosynthetic Testing for
    Waste Containment Applications held in Las Vegas on Jan. 23, 1990 and
    sponsored by ASTM Committee D-35 on Geosynthetics.
        Includes bibliographic references and index.
        "ASTM publication code number (PCN) 04-010810-38"--T.p. verso.
        ISBN 0-8031-1456-7
        1.   Sanitary landfills--Linings--Testing--Congresses.
    2.   Geomembranes--Testing--Congresses.   3. Geosynthetics--Testing-
    Geosynthetic Testing for Waste Containment Applications (1990 : Las
    Vegas, Nev.) III. ASTM Committee D-35 on Geosynthetics.   IV. Series:
    ASTM special technical publication : 1081.
    TD795.7.G45   1990
    628.4'4564--dc20   90-19311
                                                                    CIP

NOTE

The Society is not responsible, as a body,
for the statements and opinions
advanced in this publication.

Peer Review Policy

Each paper published in this volume was evaluated by three peer reviewers. The authors
addressed all of the reviewers' comments to the satisfaction of both the technical editor(s)
and the ASTM Committee on Publications.

The quality of the papers in this publication reflects not only the obvious efforts of the
authors and the technical editor(s), but also the work of these peer reviewers. The ASTM
Committee on Publications acknowledges with appreciation their dedication and contribution
of time and effort on behalf of ASTM.

Printed in Ann Arbor, MI
December 1990

Foreword

This publication, *Geosynthetic Testing for Waste Containment Applications,* contains papers presented at the symposium of the same name held in Las Vegas, NV, on January 23, 1990. The symposium was sponsored by ASTM Committee D-35 on Geosynthetics. Dr. Robert M. Koerner, Professor of Civil Engineering at Drexel University, presided as symposium chairman. He is also editor of this publication.

Contents

Overview

The purpose of the Symposium on Geosynthetic Testing for Waste Containment Applications was to focus attention on various geosynthetic testing efforts in the environmentally related application area of waste containment. In general, these containment facilities have to do with landfills, surface impoundments, and waste piles. The specific aims of the symposium were as follows;

- Assess the current state-of-the-art in geosynthetic testing as can be applied to waste containment.
- Focus upon the myriad of available geosynthetic material tests that are performed in isolation which are generally called "index" tests.
- Contrast these tests with soil/geosynthetic material tests which are generally called "performance" tests.
- Determine if, and what type of, full scale field tests are available.
- Determine if correlations exist between the above mentioned two types of laboratory tests and field performance.
- Identify needs for modification of these tests, or for additional tests which better predict field performance than those that are currently available.
- Provide a forum for information exchange among regulators and permitters, public and private owners, design and inspection consultants, testing and investigative firms, research and development organizations, manufacturers and fabricators, and installation contractors.

Slightly over half of the papers presented at the symposium and assembled in this Special Technical Publication have to do with geomembranes. This is understandable since the first line of defense of containment is a polymeric geomembrane barrier. Chemical resistance perspectives were presented in papers by Landreth, Mores et al., White and Verschoor, and Dudzik and Tisinger. These authors underscored the importance of the resistance of the base polymer with the liquid to be contained and the procedures for proper evaluation. The tests that are involved range from traditional ASTM physical and mechanical property tests on relatively large samples to chemical analysis, or fingerprinting tests, on very small samples.

Specific tests on geomembrane heat stability by Gray, transport behavior by Haxo, stress cracking by Halse et al., radiation effects by Whyatt and Farnsworth, and swelling by Thomas and Cadwallader are presented in their respective papers. These tests appear to be in a transition stage between the research mode and consideration for adoption by ASTM. In fact, some of them are in Task Group status within Committee D-35 on Geosynthetics. The data base that they bring to the literature is very welcomed. Clearly, the manner in which geomembranes are currently being challenged has reached levels of sophistication unheard of a few years ago.

Geomembrane test methods and results which are performance related were the focus of a number of papers presented at the symposium and appearing in this publication. Included in this group are durability of geomembrane seams by Peggs and Carlson, accelerated testing by Schmachtenberg and Bielefeldt, an overview of various performance tests by R. T. Sprague and Frobel, three-dimensional testing by Koerner et al., and pyramid puncture testing by Pühringer. This latter group of tests and their results should see publication in future Committee D-35 test methods and standards.

Recognizing that containment facilities are "systems" brings out the importance of other geosynthetic materials in addition to geomembranes. Within this group are geotextiles (used as filters, separators, and protection layers), geonets (used as drainage layers for leachate and other liquids), geogrids (used to reinforce cover soils or reinforce the waste itself), geocomposites (used as surface water drains) and plastic pipe, or geopipe (used for leachate removal in the primary or secondary zones of a containment system). Chemical compatibility issues regarding geotextiles, geonets, and geopipe were brought out in papers by Verschoor et al., C. J. Sprague, and Allen and Verschoor. While the incubation protocol for these materials is reasonably straightforward, the actual testing is problematic since large test specimen sizes leave little possibility for a statistical data base to be generated. The papers in this group give insight into these difficulties and possible solutions.

These papers were followed by a paper on direct shear interface friction procedures and behavior by Bove. This topic is of great importance due to several failures of cover soils sliding on geomembranes and solid waste instability in contained liner facilities. ASTM Committee D-35 has a Task Group working on this specific situation and a future test method.

The geotextile filters used in waste containment facilities were examined and challenged in a number of papers in this volume. Ling et al. and Montero and Overmann described various aspects of geotextile permeability which was counterpointed by papers on geotextile clogging by Bhatia et al., Rohde and Gribb, and Koerner and Koerner. These two aspects of filter design are all important in the geotextile's proper performance: (1) the geotextile must be sufficiently open to allow flow and (2) it must be sufficiently closed to retain the upstream soil particles. Both of these features must be accomplished without complete clogging of the geotextile filter. Involved in such clogging potential are both sediment and biological mechanisms which are described in the various papers cited.

A paper on the outdoor exposure of geotextiles by Tisinger et al. is important since the only test method currently used by the geosynthetic's community is the xenon arc accelerated weathering test and its correlation to outdoor exposure is suspect. More work on the actual behavior of geosynthetics to photo-induced degradation and its preventative measures by use of carbon black or chemical stabilizers is certainly warranted.

The concluding two papers of the symposium were on final geosynthetic cover/closure performance. These papers were by Paruvakat et al. and Levin and Hammond. The behavior of covers on solid waste landfills cannot be overemphasized. The lifetimes and proper functioning of this part of the entire system must perform for extremely long time frames. The time depends greatly on the nature and contamination potential of the underlying solid waste.

In conclusion, it is felt that the papers presented satisfied the objectives of the symposium and gave the participants an accurate and current perspective of geosynthetic testing for waste containment applications. It is hoped that this Special Technical Publication will serve the user community in a rational and logical manner as to the proper selection, testing, design, and use of geosynthetics.

Robert M. Koerner

Professor of Civil Engineering at Drexel University and Director of Geosynthetic Research Institute, Philadelphia, PA 19104; symposium chairman and editor

Chemical Resistance of Geomembranes

Robert E. Landreth

CHEMICAL RESISTANCE EVALUATION OF GEOSYNTHETICS USED IN WASTE
MANAGEMENT APPLICATIONS

REFERENCE: Landreth, R.E., "Chemical Resistance Evaluation
of Geosynthetics Used in Waste Management Applications,"
Geosynthetic Testing for Waste Containment Applications,
ASTM STP 1081, Robert M. Koerner, editor, American Society
for Testing and Materials, Philadelphia, 1990.

ABSTRACT: Geosynthetics, including membranes, geonets,
geotextiles, geogrids and plastic pipes are finding
increased use in waste management facilities. Their use
has been prompted by design requirements for improved
drainage, separation and barrier performance and economics
to provide additional free air space. An important
consideration in their use is their ability to withstand
the potential chemical attack from the waste and leachate.
This paper identifies an immersion procedure for exposing
the materials and tests for evaluating the materials after
exposure for those materials other than membranes. The
paper also discusses fingerprinting.

KEYWORDS: Geosynthetics, geonets, geotextiles, geogrids,
plastic pipes, chemical resistance, fingerprinting

Waste management facilities are being designed for active
lives of twenty years and more. The U.S. Environmental
Protection Agency (EPA) requires certain of these facilities,
hazardous waste, to perform for a 30 year post closure care
period. This means the material used in the construction of
barriers and leachate collection systems must withstand the
chemical, biological and mechanical stress over this period of
time. Geosynthetics are playing an increasing role in these
designs due to the wide availability, ease of use, efficiency of
design and economics. The EPA has previously established
procedures for evaluating membrane chemical resistance. The
method identified as method 9090(1,2) has been in use since the
early 1980's. The method has two main aspects: the immersion

Robert E. Landreth is a research engineer for the
Environmental Protection Agency Office of Research and
Development. He is the Chief of the Municipal Solid Waste and
Residuals Management Branch in the Risk Reduction Engineering
Laboratory in Cincinnati, OH 45268

conditions and the test parameters for each membrane material. The test data can be evaluated by FLEX(3), an Agency developed expert system for polyvinyl chloride (PVC), Chlorosulfonated polyethylene (CSPE) and high density polyethylene (HDPE). Since geosynthetics other than membranes are used extensively in facility designs, what procedures are recommended to review these other products? The intent of this paper is to provide initial guidance for immersion conditions and tests to be used in evaluating the exposed geosynthetic materials. The geosynthetic materials include geotextiles, geonets, geogrids, and plastic pipes. A brief discussion on the data evaluation as well as future activities will also be presented.

IMMERSION PROCEDURES

In general the immersion procedure outlined in Method 9090 should be followed for all geosynthetic materials. Method 9090 requires representative samples of the leachate from the waste management unit. The procedure requires that samples of the geosynthetic be evaluated by immersion in the stirred leachate for a period of up to 120 days. Leachate temperatures should be 23^0C and 50^0C. The containing vessel should not be of the same material as that being tested and should not compete with the geosynthetics being evaluated for potentially aggressive leachate constituents. The leachate vessel should be sealed with no free air space in order to prevent the loss of volatile constituents from the leachate. Testing of the geosynthetics should be for periods of 0, 30, 60, 90 and 120 days.

An alternative immersion procedure is being developed by the ASTM Committee D-35 on Geosynthetics and related products.

This procedure appears to be essentially the same as the EPA's Method 9090. The Agency is currently reviewing the ASTM Draft to determine its acceptability.

TEST PROCEDURES

The test procedures outlined in Table 1. are recommended. Additional tests may be required based on site-specific conditions or design requirements. It should also be recognized that the ASTM Committee D-35 is currently developing additional procedures for evaluation of geosynthetics. These new procedures may be included in the following list after their review by the Agency

Table 1. Test Methods for Geosynthetics

GEOTEXTILE:

Property (units)	Test Procedure
Thickness (mils)	ASTM D-1777 (part 32)[a]
Mass per unit area (oz/sq yd)	ASTM D-3776[a]

Property (units) cont	Test Procedure
Dimension (cm)	Direct measure machine and cross machine direction[b]
Grab tensile strength/ elongation	ASTM D-4632
Trapezoidal tear resistance	ASTM D-4533
Hydraulic burst strength	ASTM D-3786
Puncture resistance	ASTM D-3787
Permittivity[c]	ASTM D-4491
Transmissivity[d] (cm^2/sec)	ASTM D-4716[e]

GEONETS:

Property (unit)	Test Procedure
Mass per unit area	Direct measure (sample size $\geq$ 1 sq/ft)
Volatiles	Appendix G[f]
Extractables	Appendix E[f]
Thickness (mils)	ASTM D-1777 (part 32)
Rib Dimensions	Direct
Aperture size	Direct
Dimensions of configuration (cm)	Direct measure machine and cross machine direction[a]
Specific gravity or density	ASTM D-792 Method A or ASTM D-1505 (dry sample before test)
CBR puncture	GS-1[g]
Transmissivity (cm^2/sec)	ASTM D-4716[h]
Compression behavior of geonets	GN1[g]
Strip Tensile strength	Alternative method (see below)

GEO GRID:

Property (unit)	Test Procedure
Geogrid rib tensile strength	GG1[g]
Geogrid node junction strength	GG2[g]
Creep behavior and long term design load of geogrids	GG3[g]
Rib Dimension	Direct
Aperture Size	Direct

PIPE:

Property (unit)	Test Procedure
Volatiles @ 105°C	Appendix G[f]
Extractables	Appendix E[f]

Property (unit) cont	Test Procedure cont
Wall Thickness	Direct measurement (calipers)
Hardness	ASTM D-2240 Duro D
Specific gravity or density	ASTM D-792 Method A or ASTM D-1505 (dry sample before test)
Stiffness	ASTM D-2412[1]

Footnotes for Geotextiles

a   The thickness and mass per unit area measurement should be performed on all specimens before testing.

b   These measurements should be made on a single 12" x 12" specimen which has been blotted dry. The specimen should be returned to the immersion tank after measurements, i.e., same specimen for all dimension measurements.

c,d   Selection of permittivity or transmissivity should be made based on end application. If the geotextile is intended to enhance planar flow, the transmissivity test is appropriate; if for filtration, permittivity is appropriate. If the geotextile is to be used for soil reinforcement or separation only, then niether test may be appropriate.

e   The material should be aged in the waste fluid; then tested with water for a minimum of 100 hours. The profile of the geosynthetics being tested should be identical to the waste management unit. The overburden pressure should be 2-3 times the maximum expected in order to evaluate the potential for creep.

f   See reference 4.

g   These test procedures may be obtained from:

     Geosynthetic Research Institute
     Drexel University
     Philadelphia, Pennsylvania 19104
     Attention: Dr. Robert Koerner
     (215/895-2343)

h   See previous notes c,d.

i   This test requires 6" long specimens. A smaller size may be used if it can be demonstrated that results can be correlated to tests of the larger specimens. On molded pipe, care should be taken

that the test specimens all represent the same
configuration.

Alternative Strip Tensile Test Method for Geonets

The following test method is proposed as an alternative to
wide width tensile methods for evaluation of geonets. This
method should provide a satisfactory index test for monitoring
tensile properties of geonets in two directions while using a
sample conveniently sized for immersion testing.

Specimen

Rectangular specimens should be cut from parent material so
that force may be applied longitudinally in the direction of
both strands. A sample length of 5"-6" is recommended. This
sample may be mounted using standard testing machine friction
grips so that initial gauge length is 2". A minimum of three
replicate specimens should be tested for each exposure condition
and strand direction. Specimen width and grip configuration
should be selected so that load is evenly applied across no
fewer than 4-6 strands, and the specimen width should be
consistent for all tests.

Testing Proceudre

Each specimen should be pulled at a uniform crosshead speed
(2"/min is suggested). Maximum load and elongation at break
should be recorded for each specimen.

Test Report

The test report should include the following:

- Sample dimensions and number of replicates.
- Direction (i.e., inner diagonal, outer diagonal).
- Test parameters (crosshead speed, ambient temp.).
- Results (both for individual specimens and as
  average for each direction/test exposure condition).

The Agency recommends that these materials should be tested
in a "wet" condition. The "wet" condition would closely
simulate the environment in which the geosynthetic is supposed
to perform. "Wet" condition is defined as specimen removal,
rinse, blot dry, wrap (if necessary) in plastic wrap/or equal
and test. This should be performed very quickly to minimize
loss of sorbed volatiles. All specimens should be tested within
8 hours. The base line specimens should also be tested wet for
comparison testing. Test results from exposed rinsed samples
should be compared to base line samples which have been soaked
in deionized water for at least 24 hours and blotted dry before
testing; the geosynthetic should be evaluated in a condition
representative of in-situ conditions. Allowing the material to
dry could release volatile compounds that may have a detrimental

effect on the geosynthetic material. If the specimens are suspected of sorbing volatiles, head-space analysis[5] should be performed. Obviously, care should be taken to prevent contamination of laboratory test equipment and exposure of laboratory technicians.

There is also concern that geotextiles may vary greatly in weight and thickness across the roll width. Test coupons may deviate from average physical property results depending on where the sample is taken. It is suggested that the samples of geosynthetics submitted for compatibility testing should represent the full manufactured width of the material and be of sufficient length to cut the required specimens for the test. Individual specimens should be precut and randomly selected for testing. The samples that are selected for testing out of a given width should be as uniform as possible since this should not be a test of the variability of material. If the variability between specimens is large, it may be necessary to monitor the weight and thickness of each individual specimen to reduce statistical variance in the test results. This is an area where additional research is needed to clarify the issues.

FINGERPRINTING (ALL MATERIALS)

It is important to recognize that geosynthetic materials evaluated for chemical resistance be essentially identical to the geosynthetic material actually installed in the waste management unit. The Agency recommends that a "fingerprint" of the geosynthetic materials evaluated in the laboratory be compared with the geosynthetic material actually placed in the waste management unit. The purpose of fingerprinting from a chemical resistance viewpoint, is solely to identify the specific construction material that has been tested in a compatibility test. The purpose is not to identify all of the components in the compound, but to characterize the material in general terms such as:

- the polymer type and amount that is used;
- the amount of extractables; and
- the carbon black and ash contents.

It is further suggested that samples of all geosynthetic materials be retained for future verification. All specimens should be dried for 2 hours at 105°C before testing begins. The test procedures outlined in Table 2. should be used:

Table 2. Suggested Fingerpringting Techniques for Various FMLs[5]

Method *	Properties	Type of Base Resin				
		PE	PP	PET	PVC	CSPE/CPE
TGA	Composition	X	X	X	X	X
DSC	Melting Point	X	X	X	n/c	n/c
	Glass transition temp.	n/c	n/c	X	X	X
	Crystallinity	X	X	X	n/a	n/a
	Oxidation Induction Time	X	X	n/a	n/a	n/a
IR	Additives	X	X	X	X	X
	New reaction products	X	X	X	X	X
GC	Additives	n/c	n/c	n/c	X	X
	Plasticizers	n/a	n/a	n/a	X	X

Note: n/a = Not applicable
       n/c = Not common

*Methods: TGA = Thermal gravimetric analysis
           DSC = Differential scanning calorimetry
           IR = Infrared spectrometry
           GC = Gas chromatography

GEOSYNTHETIC SEAM EVALUATION

There has been some confusion as to whether geosynthetic seams should be subjected to immersion procedures of Method 9090 and tested by appropriate test procedures. It has always been the intent of the Agency that all geosynthetic seams, both factory and field, should be evaluated in a manner similar to the parent material; immersion (as per Method 9090) followed by testing. Obviously, the appropriate ASTM test procedures, depending on geosynthetic type, should be used.

The evaluation of seams after immersion (Method 9090) is only to generally qualify the seaming method. The installation QA/QC procedure should verify the adequacy of the factory and field seams. In both procedure the mode of failure for the seams should be identified in pictorial form[6]. Presentation of the data in this format should help eliminate any confusion on actual failure modes.

DATA EVALUATION

The Agency's data base on membranes is substantial. This base of information was developed as a result of a 15-year period of membrane research. This type of information is not readily available to the Agency for geosynthetics other than membranes. The Agency, under a cooperative agreement with Southwest Texas State University, Agreement No. (CR-815495), is in the process of developing the data base. The information is being solicited from manufacturers, test and evaluation firms and engineering consultants. After this base of information is reviewed, a laboratory program may be required to supplement the data base.

Fortunately, the geosynthetics in question are made from polymers similar to the polymers used in membranes. Newer materials such as polyester and polypropylene probably should react similar to the basic polymers. This allows the Agency to review these data sets in a similar manner, i.e., evaluating the rate of change in the physical properties of the material. There will, undoubtly, be some adjustment. We anticipate that the survey and laboratory program will identify where these adjustments will be necessary.

FUTURE DIRECTION

Chemical resistance remains an important consideration when selecting materials for waste management facilities. However, with the initiation of the land ban and banning of liquids, especially in hazardous waste facilities, design engineers should reevaluate their material requirements and start to design by function rather than forcing the design using one type of material.

The Agency is also considering other approaches to chemical resistance evaluation. As the concentrations of aggressive chemicals become lower the potential harmful effect on the polymers will become harder to detect. A more sensitive technique for evaluating chemical resistance needs to be developed. The Agency is aware of and reviewing the progress of several privately sponsored research areas such as thermal alnalytical techniques and individual fibers versus bulk samples.

SUMMARY

A review of the physical properties after immersion in the waste liquid has been and continues to be, the method by which the Agency has approved and disapproved chemical resistance for geosynthetics. Immersion procedures are described that are essentially identical for all geosynthetics. Test methods for geotextiles, geonets, geo grids and plastic pipe have been presented. A discussion on "wet" testing and "fingerprinting" now sets forth a consistant approach for evaluating all geosynthetics for chemical resistance. Realizing that changes are occurring within the waste management industry, a look at the future is also presented.

REFERENCES

[1]   U.S. Environmental Protection Agency. EPA Method 9090,
        Compatibility Tests for Wastes and Membrane Liners. In:
        Test Methods for Evaluating Solid Waste, Third Edition.
        EPA SW-846, U.S. Environmental Protection Agency,
        Washington, D.C., 1986.

[2]   U.S. EPA, Office of Solid Waste and Emergency Response.
        U.S. EPA 1986 Supplementary Guidance on Determining
        Liner/Leachate Collection Compatibility. U.S.
        Environmental Protection Agency, OSWER Directive 9480.00-
        13. Washington, D.C., 1986.

[3]   Landreth, Robert E. FLEX - An Expert System to Assess
        Flexible Membrane Liner Materials. In: Proceedings: 89
        Geosynthetics Conference, Industrial Fabrics Associates
        International, St. Paul, MN, 1989.

[4   U.S. EPA Lining of Waste Containment and Other
        Impoundment Facilities. EPA/600/2-88/052, U.S.
        Environmental Protection Agency, Cincinnati, OH, 1988.

[5]   ibid, Section 4.2.2.5.1, pp 4-88 - 4-107.

[6]   ibid, Appendix N.

Maryanne Mores, Patrick E. Cassidy, David J. Kerwick and Deborah C. Koeck

A REVIEW OF POLYMER TEST METHODS APPLICABLE TO GEOSYNTHETICS FOR WASTE CONTAINMENT

REFERENCE: Mores, M., Cassidy, P. E., Kerwick, D. J., and Koeck, D. C., "A Review of Polymer Test Methods Applicable to Geosynthetics for Waste Containment," Geosynthetic Testing for Waste Containment Applications, ASTM STP 1081, R. M. Koerner, Ed., American Society for Testing and Materials, Philadelphia, 1990

ABSTRACT: The study of the environmental degradation of geosynthetic materials used in hazardous waste facilities has generated much interest in recent years. Analytical methods are used to supplement conventional structure changes and mechanisms of failure associated with the polymer. Any changes in the base polymer caused by chemical exposure may result in severe changes in strength, durability and serviceability of the geosynthetic. Chemical and physical analytical methods are used in evaluation of chemical compatibility and long-term durability of geosynthetics for several purposes: 1) to verify that geosynthetics tested in advance of landfill construction are essentially the same as those procured for the actual installation, 2) to identify and monitor changes in molecular structure caused by chemical attack, and 3) to provide accelerated assessment of the effects of long-term aging, an especially difficult and as yet, not thoroughly researched topic. These factors can be reviewed by studying the oxidation and solvation properties as well as the change in crystalline structure of the plastics.

This paper discusses "fingerprinting" methods used to characterize geosynthetics. These methods fall into four major categories of analytical techniques: thermal analysis, spectroscopy, chromatography and microscopy. The advantages of these analytical techniques are their precision, accuracy, and small sample size requirements. The methods discussed in this paper can detect small changes in molecular structure that often are not observed with traditional bulk properties testing after short aging times.

KEYWORDS: Analytical techniques, geosynthetic characterization, spectroscopy, thermal analysis, microscopy, chromatography

Dr. Cassidy is a professor of chemistry at Southwest Texas State University (SWTSU) and principal investigator of a study sponsored by the USEPA to investigate the chemical compatibility of geosynthetics other than geomembranes. M. Mores is the project manager and D. Kerwick is a graduate student at SWTSU. D. Koeck, a former graduate student at SWTSU, is currently at Texas Tech University. Correspondence may be directed to M. Mores, Department of Chemistry, SWTSU, San Marcos, TX 78666.

INTRODUCTION

Over the last ten to fifteen years, the disposal of hazardous and municipal wastes has become one the most significant and controversial problems facing our society. Although recycling, incineration and waste reduction have partially addressed this problem it is inevitable that a certain amount of these wastes would be stored in the ground. Geosynthetics, polymeric materials used in a construction role, have been instrumental in the development of waste containment technology. Though these materials offered many advantages over traditional construction materials, they lacked extensive prior use and, therefore, very few data or technology exist for evaluating their performance and almost none for long-term life evaluation. Federal, state and local governments have devised stringent permitting procedures to help establish the performance of geosynthetics.

Early practitioners in this field chose evaluation techniques which had been used to characterize more traditional construction materials. These tests were very performance oriented and were excellent in describing current performance. The ability to predict long term performance especially against chemically aggressive media is not adequately covered by these methods.

Recently, the use of analytical methods to assist in characterizing geosynthetic materials' performance has begun. Polymer chemists have for years developed and utilized analytical test methods to determine the performance of polymeric materials in an aggressive chemically-degrading environment. The approach developed was to look at mechanics of degradation at a molecular level. This requires test methods which will give information about a variety of properties to include molecular structure, composition, thermal and optical properties. By the study of these properties and their changes after a chemical interaction has occurred, information can be gained that will help predict the performance of geosynthetic materials. This review will discuss some of these methods, their applications and the information gained from each.

STRUCTURE OF BASE POLYMER RESINS

The durability of geosynthetic materials is dependent to a great extent upon the composition of the polymers from which they are made. To quantify the properties of these materials, a knowledge of their structures on the chemical, molecular and supermolecular level is necessary.

Chemical Structure

The chemical composition and structure of the polymer from which the geosynthetic is made is directly related to the properties that the finished materials exhibit. These structure-property relationships allow manufacturers to custom-design polymeric materials for specific applications.

Polyethylene (PE), the simplest organic polymer, has the least reactive chemical structure of all commercial thermoplastics. PE produces widely-used geosynthetic materials due to this lack of reactivity, ease of processing, low cost and acceptable physical properties.

Polypropylene (PP), also devoid of relatively reactive functional groups, is used only in its isotactic, stereoregular form. This stereoregular form allows the molecule to form a symmetric helical structure that can pack into oriented crystalline units and, therefore, be a relatively tough material. It is similar to polyethylene, but possesses better flexural properties.

Polyester, in its most common form, poly(ethylene terephthalate) (PET), is a condensation polymer of a dibasic acid and a dialcohol. The ester group, $--CO_2--$, the important polymeric link, can be hydrolyzed under certain alkaline conditions. The planar aromatic group stiffens the polymer backbone and decreases the inherent flexibility of PET, as compared to aliphatic polyesters.[1-4]

Polyvinyl chloride (PVC) is a linear addition polymer that is used commonly with fillers or plasticizers. Rigid PVC, as used in pipe, is a filled, unplasticized material while flexible applications require the polymer to be compounded with plasticizers (oils or waxes). PVC is known for its ability to absorb organic liquids and to consequently experience a softening (lower T_g). It is also susceptible to ultraviolet light- or heat-induced degradation to become brittle and darken.

Molecular Structure

The degree of polymerization or average molecular weight of polymers refers to the average number of monomer units in the chain or its length. Molecular weights of the polyolefins, PE and PP, which are produced by addition polymerizations, can range from 200,000 to 1,000,000. Polyesters, produced from condensation reactions, usually produce molecular weights in the 10,000 to 30,000 range. The polyolefins require higher molecular weights due to the absence of functional groups which would provide inherent chain stiffening to allow control of chain flexibility. The physical and chemical properties of the polymers used in geosynthetic materials require sufficiently high molecular weight; however, too high a molecular weight will lead to processing problems.[1-4]

The sequencing of asymmetric or chiral carbons which contain a substituent group at each monomer residue refers to the stereoregularity or tacticity of the polymer. This capacity to form stereoregular polymers results in three primary possibilities which are characterized by nuclear magnetic resonance (NMR) spectroscopy. Isotactic (Figure 1a) sequences have the same configuration along the polymer backbone; syndiotactic (Figure 1b) sequences exhibit pendant groups about the chiral center arranged in an alternating manner; and atactic (Figure 1c) sequences demonstrate a random or irregular arrangement along the backbone. Tacticity is determined by the bond making process during polymerization and is solely dependent on polymerization conditions and catalyst type.[5] This occurrence of stereochemical configurational isomerism markedly affects the bulk physical properties. Isotactic and syndiotactic isomers result in high-melting compounds that crystallize readily, whereas atactic sequences produce materials that do not crystallize readily and are low-melting or even rubbery.[6,7]

Due to the random nature of polymerization involving synthetic polymers, a mixture of polymer chains with different molecular weights (broad molecular weight distribution , MWD) is produced, and referred to as polydispersity. Physical properties of polymer products such as melt viscosity, tensile strength, modulus, impact strength or toughness, and resistance to heat and corrosives are dependent on the molecular weight and the MWD.[7] The size distribution is described quantitatively by the MWD function. The distribution may be established either by determination of a sufficient number of the various molecular weight averages or by fractionation by gel permeation chromatography (GPC). [5]

Supermolecular Structure

The three dimensional structure or fine structure of macromolecules is important for all geosynthetics, but particularly so for fiber-forming polymers used to make geotextiles. The manner in which the polymer organizes to form well-oriented, closely-packed crystalline regions and randomly-coiled, amorphous regions must be properly balanced to produce the correct physical properties necessary for fibers used in geotextiles.[1-4, 8]

Polymers crystallize, to a limited extent, only if molecular structure is ordered, whereas, non-polymeric solids are essentially 100% crystalline. In polymeric compounds, the remaining noncrystalline portion constitutes randomly-coiled, disordered (amorphous) chains. Polymers with structurally regular chains, such as high density polyethylene (HDPE), possess the ability to disentangle and form crystals. Polymer crystallization depends on the presence or absence of tacticity and the minimum-energy conformation of the chain.[6] As the degree of crystallinity increases, structural irregularity decreases. Thus, compounds that suggest highly ordered, i.e. isotactic or syndiotactic, sequences with minimal branching are highly crystalline. Crystallinity has a profound effect on polymer properties, especially mechanical properties, due to the smaller polymeric crystals in which the evenly and tightly packed molecules result in high intermolecular forces and dense regions. Therefore, an increase in the degree of crystallinity directly correlates to an increase in modulus, stiffness, yield and tensile strength, hardness and softening point and a decrease in permeability.[5]

Microcrystalline polymers maintain a regular packing of chains in small domains in a matrix of amorphous polymer. This type of orientation results in a higher density than polymers that are totally amorphous. Microcrystalline regions held together by dipolar, hydrogen bonding, chain entanglement or van der Waals forces provide what could be called "physical crosslinks" for the amorphous regions. (The chains do not have primary chemical interchain bonds.) When introduced to an amorphous polymer, microcrystalline crosslinks stiffen and toughen the polymer and a rubbery, elastomeric polymer changes into a tough flexible material. Microcrystalline polymers bend better without breaking, resist impact better and are less affected by temperature changes or solvent penetration than completely amorphous polymers.[6]   However, except in those cases where branching is purposely introduced to inhibit crystal formation and lower density, increased branching on the backbone is undesirable since branching disrupts crystal formation. Orientation (such as cold drawing) of crystalline polymers results in improved physical properties when fibers are stretched and/or processed. As fibers stretch, the molecules become more oriented and tend to crystallize, and in turn, are stronger, tougher and somewhat more elastic than unoriented fibers. High molecular symmetry and high cohesive energies between chains, which both require a fairly high degree of polymer crystallinity result in high tensile strength and high modulus in fibers. Order also increases when polymer films are biaxially oriented.[7] At the molecular level, the crystal structure is characterized by wide angle x-ray diffraction.

The properties of geosynthetic materials are directly related to the structure of the polymer from which the materials are made. The chemical properties of the base polymer, the average molecular weight, the molecular weight distribution, and the crystallinity combine to contribute to the chemical and environmental durability of the geosynthetic.[1-4, 8]

GEOSYNTHETIC CHARACTERIZATION METHODS

No standardized methods exist for analyzing the durability of geosynthetic materials before or after exposure to the environment or aggressive chemicals. Microstructural techniques exist and aid in the determination of changes in the molecular structures of the base polymers used to manufacture the geosynthetic. In the past, the durabilities of the geosynthetics have traditionally been assessed on the basis of mechanical property test results, not on the microstructural changes that cause the changes in the mechanical properties. Specialized analytical techniques examine the molecular structure of geosynthetic materials, and elucidate the relationship between microstructural changes in the base polymer and physical and mechanical property changes of geosynthetics.[9] They comprise analytical methods commonly used to characterize components on the molecular level.

Thermal Analysis

Thermal analysis includes a range of techniques for determining the temperature dependence of the polymer property changes. Mass, heat capacity, mechanical response and volume are the most commonly observed physical property changes examined by thermal analytical methods. A good indicator of structure-behavior relationships in polymer applications, this measurement of property changes *versus* temperature has prompted the development of these techniques specially for polymer work.[5]

Differential scanning calorimetry: Differential scanning calorimetry (DSC) is a thermal analytical technique in which the difference in the amount of heat absorbed or emitted by a polymer sample is measured by the power consumed as the temperature is increased.[7] As a transition occurs in the sample, thermal energy is added to the sample or reference material (an inert standard) to maintain the same temperature for both. This energy, which is recorded, compensates for that lost or gained as a consequence of endothermic (absorbed energy) or exothermic (emitted energy) reactions taking place in the sample. This provides a direct calorimetric measurement of the transition occurring in the polymer (glass transition or crystalline melting point).[10,11] The resulting thermal curve displays energy (MJ/sec or Mcal/sec) as a function of temperature (°C). Endotherms provide information used to calculate melting point ranges and the degree of crystallinity. Exotherms allow the assessment of the oxidative stability of the material based on the oxidative induction time or the oxidative induction temperature.[12, 13] DSC also measures decomposition onset temperatures.

Thermogravimetry: Thermogravimetry (TG) furnishes information about the composition of a geosynthetic as elucidated by its thermal degradation. TG involves gradual heating of the sample in an inert atmosphere or air to temperatures of as much as 1000°C while monitoring the weight loss as a function of temperature. The weight loss corresponds to the volatilization of various components of the test sample, there being additives or products of decomposition of the actual polymer. Data commonly derived from TG are concentration of additives, polymer, carbon black, ash, and onset of decomposition temperatures.[12, 13]

Thermomechanical analysis: Thermomechanical analysis (TMA) is an analytical technique in which the penetration, expansion, contraction or extension of a substance is measured as a function of temperature. This technique measures the deformation of the material under nonoscillatory compressive load as it is subjected to a controlled temperature program.[10, 14] Interchangeable sample probes are used to determine these changes in the materials being tested. Polymer swelling and dissolution in liquids can be detected by TMA operated in its isothermal mode. Information may be readily obtained for a variety of solvent swelling agent systems that usually require more sophisticated experimental techniques.[15]

Softening, heat distortion and glass transition temperatures may be detected with a small-tip diameter probe and a loaded weight tray. Coefficients of thermal expansion and dimensional changes due to stress relief may be examined by probes of large tip diameter and zero loading in the expansion mode. Samples may be tested in the form of plugs, films, pelletized powders or fibers. The temperature range falls between that of liquid nitrogen and 850 °C.

Dynamic mechanical analysis: Dynamic mechanical analysis (DMA) determines the mechanical properties of materials as they are deformed under periodic stresses. This most sensitive of thermal analytical techniques measures the resonant frequency and mechanical damping as a function of temperature. The sample (a thin film) is subjected to a controlled temperature program as it is forced to flex at a selected amplitude. The temperature range at which samples are tested is -150 °C to 500 °C. DMA detects transitions associated with the movement of polymer chains. This method generates information about the viscoelastic properties of a material (usually semicrystalline) such as storage modulus, loss modulus, stress relaxation and creep compliance. DMA is helpful in determining the effectiveness of reinforcing agents and fillers used in thermoset resins.[10, 14]

Spectroscopy

Spectroscopic techniques provide information on the compositional and structural characteristics of a polymeric material. Infrared (IR) spectroscopy is particularly useful in the identification of characteristic functional groups and molecular configurations by simple inspection and reference comparison. The sample is subjected to infrared radiation of successively decreasing frequencies. A series of spectral bands is generated, each of which correlates to a particular frequency or range of frequencies where the organic functional group in the plastic material absorbs radiation. These characteristic absorption frequencies indicate certain functional groups in the polymer.[9] Comparative IR techniques can be a sensitive method to detect changes in the polymer.

Fourier transform infrared spectroscopy: Fourier transform has been applied to various spectroscopic methods to enhance the sensitivity of these methods using a high speed computer that isolates very weak signals from environmental noise.[11] Fourier transform infrared (FTIR) spectroscopy involves the splitting of the incident beam into waves followed by their recombination. Spectral information is obtained from the phase difference between these two waves by a Michelson interferometer. The information is digitized, transformed mathematically from the time domain and converted to a conventional infrared spectrum.[5] Special advantages in FTIR spectroscopy include energy limited, time limited or signal-to-noise limited situations. FTIR has the ability to look at intractable, thick and intensely absorbing materials which has lead to its particularly wide application in polymer characterization. In addition, this technique observes physical and chemical changes in the polymer structure as they occur.[16] This method enhances the capability to evaluate ongoing chemical and physical property changes of polymeric materials on a small scale upon exposure to hazardous materials.

Attenuated total reflectance Fourier transform infrared spectroscopy: Attenuated total reflectance (ATR) FTIR spectroscopy provides information on the composition of complex samples and the surface chemistry of polymeric materials. The sample is placed in contact with a transmitting prism which is then fixed in the sample compartment. The distance to which the infrared radiation appears to penetrate in internal reflection depends on the wavelength, but is of the order of 5 mm. It is possible, by scanning over a range of infrared frequencies, to correlate specific vibrational absorption maxima with the atomic groups responsible for the absorptions. These correlations provide a powerful tool for the identification of particular covalent bonds. If the surface of the polymeric material has been degraded upon exposure to hostile chemicals, this technique gives information regarding changes in surface chemistry resulting from the exposure.[16, 17]

Chromatography

Chromatographic methods allow the separation, isolation and identification of closely related components in complex mixtures. Due to the diversity of these separation techniques, other analytical methods do not possess the ability to identify components from mixtures with such accuracy. Chromatography identifies components in the gaseous, liquid or solid state, which may include substances that have been absorbed into the geosynthetic material itself. This valuable information can be used to identify various components of leachates or products that result from polymer degradation.

Gel permeation chromatography: Gel permeation chromatography (GPC) is an extremely valuable analytical technique for determination of molecular weight distribution. Used with a variety of solvents (tetrahydrofuran, benzene, xylene, chloroform, dimethylformamide or fluorinated alcohols) for a variety of polymer systems, GPC requires only small samples (typically 0.5 to 3 ml of a 0.05 to 0.1% solution of polymer). The technique utilizes a rigid porous column (1 to 3 m, or 3 to 10 feet, packed with highly crosslinked, porous, solvent-swelled polystyrene). The degree of swelling of the stationary phase determines the resolution of the separation. The column works as a sieve with the larger, higher molecular weight polymers passing through the gel column more quickly than the smaller, lower molecular weight molecules due to the retention of the latter in the pores of the stationary phase.[6, 18] A detector is then used to determine the difference in refractive index (differential refractometer) or absorbance (ultra-violet spectrophotometer), which relate to the amount of polymer in the eluted solvent, as a function of time to produce the molecular weight distribution.

Gas chromatography: Gas chromatography (GC) fractionates the components of a vaporized sample as a consequence of partition between a mobile gaseous phase and a stationary phase held in a column.[19] Headspace gas chromatography is an important analytical technique for qualitative and quantitative determination of volatile components of complex samples, even in minute concentrations.[20-22] The method requires vaporization of the volatiles in a confined headspace and analysis of the resulting gas by gas chromatography. Separation of the sample is achieved by elution using an inert carrier gas (helium) and a stationary solid phase that has different affinities for the sample components. Results are received in graphical form displaying peaks with varying retention times (identifying the component) and peak area (from which amount is calculated).[11]

Microscopic Analysis

The use of magnification to evaluate geosynthetic materials is provided by various types of microscopic techniques. Microscopy supplies rapid, direct observational information about the surface as well as the internal microstructure and defect distribution within the materials. Microscopic analysis is also a valuable tool for the examination of geosynthetic materials before and after exposure to aggressive chemical media.

Optical microscopic analysis: Cross sections of polymeric materials are examined using microtome specimen samples or in an innovative technique utilized by Rollin,[23] where optical fibers are used to carry the light directly in to the geosynthetic. Cross-sections can be observed and photographed using minimum resolution (100 to 200 nm) and magnifications of up to 2000 times. Flow patterns of the polymer, microcracks and air channels can be found in substantially less time and with greater precision than before realized.

According to Peggs & Charron,[24] the microtome technique of microstructural examination has been found to contribute to the development of new and modified materials, and subsequent processing methods. The distribution and relative magnitude of residual stresses can be identified by crossed or partially crossed polarizing filters. Characteristics of fracture phenomena can also be determined during failure analysis. Microstructural evaluation of the basic geomembrane facilitates the improvement of the standard method of determining the dispersion of carbon black in polyethylene geomembranes.

Scanning electron microscopy: A powerful and versatile analytical tool for determining microstructural changes in geosynthetic materials, the scanning electron microscope (SEM) has resolution limits of 4 to 5 nm and magnifications of up to 60,000. Microcracks that undermine the strength and durability of geosynthetics are easily visible within the analytical parameters of SEM.[23]

CHEMICAL EXPOSURE AND DEGRADATION

Geosynthetics used in hazardous or municipal waste containment will be exposed to numerous potentially degrading chemicals (acid, base, various organics). The interaction that takes place will depend both on the degrading chemical and the base polymer from which the geosynthetic is made. Exposure to an alkaline solution may have an entirely different effect of polyethylene terephthalate than it does with polypropylene.[25] Temperature may also play a role in the kind or amount of degradation that occurs. Use cf analytical methods will allow these interactions to be more fully understood.

CONCLUSION

The use of analytical methods in the characterization of geosynthetics provides data from the molecular level. Coupled with information gained in more traditional mechanical testing, a better picture of the performance of these materials can be understood. "Fingerprinting" analysis will aid in quality control and quality assurance. Studies of chemical structure, molecular weight, tacticity and superstructure all provide information unavailable through more traditional test methods. Monitoring changes in the above as part of an accelerated aging program should allow for earlier detection of small changes in the properties of the geosynthetic material. Through augmentation rather than replacement of mechanical and hydraulic test methods, analytical test should find increasingly further use in future studies of geosynthetic materials.

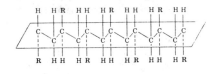

(a)

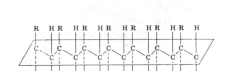

(b)

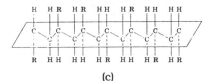

(c)

Figure 1. Diagrams showing a) the stereoregular syndiotactic, b) isotactic and c) the irregular atactic configurations in a vinyl polymer.[18]

ACKNOWLEDGEMENTS

The authors appreciate the support for this project by the USEPA (Grant No. GR 815495). The authors would also like to thank Robert M. Koerner, Drexel University, Philadelphia, PA, Karen L. Verschoor, David F. White, Rock Rushing and Rick Thomas, Texas Research Institute, Austin, TX for their comments and helpful suggestions. We are also grateful to Keith Brewer and Kurt French, SWTSU, for preparation of this manuscript and literature compilation.

REFERENCES

[1] Rebenfeld, L., Cooke, T., "Structure and Properties of Fibers in Relation to Durability of Geotextiles", a report by TRI-Princeton, Princeton, NJ, 1989.

[2] Rollin, A. L., Lombard, G., Geotextiles and Geomembranes, Vol. 7, 1988, pp. 119-145.

[3] Cooke, T., Rebenfeld, L., Geotextiles and Geomembranes, Vol. 7, 1988, pp. 7-22.

[4] Rebenfeld, L., "Chemical and Physical Structure of Fibers in Relation to the Durability of Geotextiles", Proceedings of a Seminar on Durability and Aging of Geosynthetics, Dec. 8-9, Philadelphia, PA,1988.

[5] Alger, M. S., "Polymer Science Dictionary", Elsevier Science Publishers Ltd., Essex, England, 1989.

[6] Allcock, H. R., Lampe, F. W., "Contemporary Polymer Chemistry", Prentice Hall, Inc., Englewood Cliffs, NJ, 1981.

[7] Seymour, R. B., Carraher, C. E., Jr., "Polymer Chemistry: An Introduction", Marcel Dekker, Inc., New York, NY, 1981.

[8] Daniel, J. L., "Is Microstructure Important to the Performance of Geosynthetics?", ASTM Symposium on Microstructure and the Performance of Geosynthetics, Orlando, FL, January 27, 1989.

[9] Tisinger, L. G., "Microstructural Analysis of the Durability of a Polypropylene Geotextile", Geosynthetics '89 Conference Proceedings, Feb. 21-23, San Diego, CA, 1989, pp. 513-524.

[10] Willard, H. H., Merritt, L. L., Dean, J. A., Settle, F. A. "Instrumental Methods of Analysis", Wadsworth Publishing Co., Belmont, CA, 1981.

[11] Skoog, D. A., "Principles of Instrumental Analysis", Saunders College Publishing, Philadelphia, PA, 1985.

[12] Thomas, R. W., Verschoor, K. L., Geotechnical Fabrics Report, May/June, 1988, pp. 12-15.

[13] Thomas, R. W. , Verschoor, K. L., "Thermal Analysis of Nonwoven Polyester Geotextiles," a report by Texas Research Institute, Austin, TX, 1988.

[14] Wendlandt, W. W., Gallagher, P. K. in "Thermal Characterization of Polymeric Materials," Turi, E., Ed., Academic Press, Inc., Orlando, FL, 1981, pp. 68-72.

[15] Machin, D., Rogers, C. E., Polymer Engineering and Science, 10 (5), 1970, pp. 300-304.

[16] Grasselli, J. E., Mocadlo, S. E., Mooney, J. R. in "Applications in Polymer Synthesis and Characterization: Recent Developments in Techniques, Instrumentation, Problem Solving", Mitchell, J., Ed., Hanser Publishers, New York, NY, 1987, pp. 316-319.

[17] Rushing, R., Thomas, R. W., Private Communication, 1989.

[18] Billmeyer, F. W., "Textbook of Polymer Science", Wiley Interscience, New York, NY, 1971.

[19] Skoog, D. A., West, D. M., "Fundamentals of Analytical Chemistry", Saunders College Publishing, New York, NY, 1982.

[20] Ettre, L. S., Kolb, B., Hurt, S. G., American Laboratory, 15 (10), 1983, 76-83.

[21] USEPA, "Method 8.82, Headspace Method. Test Method for Evaluating Solid Waste: Physical/Chemical Methods", SW-846, Office of Solid Waste USEPA Washington, DC, 1980.

[22] Lattimer, R. P., Pausch, J. B., American Laboratory, 12 (8), 1980, pp. 80-88.

[23] Rollin, A. L., Vidovic, A., Denis, R. and Marcotte, M., "Evaluation of HDPE Geomembrane Field Welding Techniques: Need to Improve Reliability of Quality Seams", Geosynthetics '89 Conference Proceedings, Feb. 21-23, San Diego, CA, 1989, pp. 443-455.

[24] Peggs, I. A., Charron, R. M., "Microtome Sections for Examining Polyethylene Geosynthetic Microstructures and Carbon Black Dispersion", Geosynthetics '89 Conference Proceedings, Feb. 21-23, San Diego, CA, 1989, pp. 421-432.

[25] Koerner, R. M., Lord, A. E. Halse, Y. H., Geotextiles and Geomembranes, Vol. 7, 1988, pp. 147-158.

David F. White and Karen L. Verschoor

PRACTICAL ASPECTS OF EVALUATING THE CHEMICAL
COMPATIBILITY OF GEOMEMBRANES FOR WASTE
CONTAINMENT APPLICATIONS

REFERENCE: White, D.F., and Verschoor, K.L., "Practical Aspects of
Evaluating the Chemical Compatibility of Geomembranes for Waste
Containment Applications," Geosynthetic Testing for Waste
Containment Applications, ASTM STP 1081, Robert M. Koerner,
editor, American Society for Testing and Materials, Philadelphia, 1990.

ABSTRACT: Chemical compatibility testing (EPA Method 9090) has
been applied to many combinations of natural and synthetic leachates
and geomembranes. A number of commercial laboratories are
currently performing the method, and significant experience has been
gained in its application. This paper will describe some practical
aspects of chemical compatibility testing which have implications for
proper selection of materials for hazardous waste containment.

KEYWORDS: Chemical compatibility, geomembrane, liner, waste
containment, EPA Method 9090.

Introduction

Since the onset of regulations requiring verification of chemical
compatibility for flexible membrane liners, many combinations of natural
and synthetic leachates have been tested against proposed liner materials
under procedures specified in EPA Method 9090 [1]. A number of
commercial laboratories are currently providing these testing services, and
significant experience has been gained in applying the method. The
following problem areas are discussed in this article:

- Test objectives and purpose or rationale for performing the test;

- Failure criteria and test variability;

- Stress-strain measurement and modulus of elasticity for HDPE
  liners;

Mr. White is Materials Laboratory Manager at TRI Environmental, Inc., 9063 Bee
Caves Road, Austin, TX 78733; Ms. Verschoor is Program Manager for Geosynthetics at
TRI Environmental, Inc.

- Volatiles and extractables determination for HDPE;

- Quality control of laboratory data; and

- Cost.

Recommendations are provided which may result in more meaningful evaluation and interpretation of test results.

Review of Method 9090

Geomembranes are increasingly being specified for use in containment facilities for both hazardous and municipal wastes. Materials specified for waste containment facility construction must be resistant to (or compatible with) the liquids they will contact during the lifetime of the facility. The US Environmental Protection Agency (EPA) has provided standards and additional guidance to assure the facility design meet this criterion, and to provide some assistance for interpretation of complex federal legislation governing landfill design and construction. The risk to the environment is great should a landfill cell fail due to degradation of synthetic construction materials.

Performance of each component depends on long-term stability of mechanical and hydraulic properties. Degradation caused by interaction with the waste stream can reduce the ability of a liner material to retain its original design function. This degradation may be independent of other considerations such as static compressive loading and settlement, natural material aging, or biological growth. Chemical compatibility testing is required to provide documentation for the facility operating permit application.

EPA Method 9090 was developed in the early 1980s to provide a means to determine the compatibility of geomembranes with waste liquids by stimulating some of the conditions a geomembrane would experience in a waste management facility. Rectangular geomembrane samples are immersed in a chemical environment representative of the waste liquids or leachates to be contained for minimum periods of 120 days at room temperature (22°C) and at elevated temperature (50°C). Samples are immersed in exposure tanks designed to support the samples so that they do not touch the bottom of the tank or each other and to prevent the loss of volatile leachate components. Various physical properties of the geomembrane are monitored every 30 days. Physical properties measures before and after immersion are compared to evaluate the compatibility of the selected with the representative exposure media or leachate.

EPA Method 9090 requires that all geomembranes be tested for hardness, tensile properties, puncture resistance, specific gravity, volatiles content, extractable content, hydrostatic resistance, dimensions and weight.

In addition to these properties, geomembranes may be tested for the following according to construction:

Semicrystalline Geomembranes	Fabric-Reinforced Geomembranes
Modulus of Elasticity Tear Resistance	Ply Adhesion

The most recent revision of EPA Method 9090 suggests that semicrystalline geomembranes be tested for environmental stress cracking. It is also suggested that unexposed and exposed field seams be tested for seam strength in the peel and shear modes.

Test Objectives and Rationale for Performing the Test

Some misconceptions exist about the reason the Method 9090 was developed, and the objectives it was designed to meet. Such questions as the following have arisen:

- Why consider a temperature like 50°C? No liner would be exposed to such a high temperature in service. Or, the temperature is too low; the test is not severe enough to provide a significant degree of time compression.

- Why are only two temperatures included? If more temperature levels were evaluated, an Arrhenius rate constant might be derived, making possible a true predication of life.

- Why use a maximum test time of 120 days leachate exposure?

- Why are so many different properties measured?

Overall design of the test protocol was based on long term aging studies of synthetic liner materials performed by Matrecon, Inc. under USEPA sponsorship in the 1970s. Method 9090 evolved as a compromise between the desire to produce relatively long term performance data for studies lasting one or two years or more, and the recognized need to develop a standardized procedure which could be completed in less time as part of the Resource Conservation and Recovery Act permitting process.

The real value of the test lies in its ability to detect a process in which a hazardous waste interacts with a polymeric liner material, resulting in change in physical properties. The test was never intended to be used as a true life prediction tool, and the resulting data is not sufficient to support a statistical prediction of life based on reaction rate theory for the following reasons: (1) 120-day exposures are not long enough, and (2) the multiple temperatures required to produce Arrhenius shift factor curve are not required. Life prediction requires a number of assumptions which are impossible to verify:

- Degradation is caused solely by a heat-activated chemical reaction, the rate of which may be measured.

- The reaction mechanism is the same at all test temperatures, and at the service temperature.

- The leachate tested is identical in composition to that found at the site, throughout the life of the liner.

- Service temperatures and environmental conditions will remain constant and in agreement with predictions throughout the life of the facility.

- Eventual liner failure will result from aging-induced chemical structure degradation, not factors such as installation stresses, improper seaming or environmental stress cracking.

Clearly, a life prediction model which makes all these assumptions would be difficult to defend for a real waste containment facility. This fact, together with the difficulty in validating results of service life prediction models without historical data, makes true life prediction an extremely difficult proposition.

Method 9090 was designed to apply to several classes of materials, including both supported and unsupported sheet liners incorporating thermoplastic or thermosetting resins or elastomers. The material classifications in the method require different types of physical property tests, depending on the type of liner. Although a large volume of testing is required, measurement of several physical properties helps in data interpretation, since each of the physical properties are subject to significant variability in testing.

The temperatures and exposure times selected were somewhat arbitrary, and reflect the need for a test which is not prohibitively expensive and will not delay a construction project any more than absolutely necessary. The 120 day exposure time was a compromise, since it was felt that longer times are certainly desirable to verify long-term performance. Exposure for 120 day allows four 30-day interval points for comparison of test values. There is general agreement that exposure at elevated temperature will accelerate potential degradation mechanisms. 50° C is a temperature which is outside the expected service conditions for almost any liner installation, but well below melt or degradation temperatures for any of the common polymers from which liners have been made.

Failure Criteria and Test Variability

Data for EPA Method 9090 are commonly presented to the permit reviewer to demonstrate the chemical compatibility of a selected geomembrane. The permit reviewer must evaluate the results for this testing and decide whether the proposed geomembrane is in fact chemically resistant.

Evaluating chemical compatibility data is a complex task that requires insight as to how the tests were run, whether the character of the exposure media has changed during the period of testing, and the variability inherent in the unexposed geosynthetic material. For example, it is not unusual for HDPE geomembranes to vary in ultimate tensile strength by as much as 30 percent between individual, baseline test coupons. It becomes very difficult to say that a 20 percent reduction in strength measured after chemical exposure is due to degradation when the baseline results have a much larger range among replicates. Statistical measures of central tendency must be reported and evaluated to assess the significance of any apparent change after chemical exposure. It is also important to look at general trends and consistency considering all test results cumulatively. To apply an objective pass/fail criterion to one individual index property is to lose perspective of the overall analysis. If the materials were in fact degrading, one should expect a corresponding loss in other physical properties. '

Another concern with evaluating chemical compatibility data is that the observed trends may be due to a reversible effect: solvent swelling. it is common for polymer materials to absorb organic chemicals and, after retaining these chemicals, to soften and swell. It is not unusual for these materials to regain their original properties after the chemicals have been removed from the polymer matrix. Absorption and desorption in this manner are governed by the laws of diffusion. A strength loss due to solvent absorption will reach a minimum value with time and may still be within the design constraints of the facility.

High-density polyethylene (HDPE)ia a highly inert material. Absorption-swelling phenomena, not irreversible chemical-induced degradation, are the predominant form of interaction with test leachates. Swelling is evidenced in test results by:

- Slight weight increases (<2% typically);

- Decreases in elastic modulus and tensile stress at yield; and

- Increases in elongation at yield and break.

Table 1 illustrates this effect, using data from an actual exposure of HDPE geomembrane to a waste leachate. In this example, the effects listed above are consistent within tensile properties measured throughout

TABLE 1.   TENSILE PROPERTIES OF A LINER EXPOSED TO A
CHEMICAL LEACHATE AT 50 C

PROPERTY	O DAY AVG	30 DAY AVG	60 DAY AVG	90 DAY AVG	120 DAY AVG
Elongation at Yield (%)	15	27	28	30	20
Tensile Stress at Yield (psi)	3210	2270	2220	2400	2590
Elongation at Break (%)	890	980	1000	950	930
Tensile Strength at Break (psi)	4440	4020	4080	4210	4120
Modulus of Elasticity (psi)	69500	40100	29300	31400	38000

NOTE: MEAN VALUES FOR FIVE SPECIMENS

the 120-day exposure period. Note especially that the degradation process is not a monotonic or continuous function; the properties shift through the 30 and 60-day time intervals, and remain relatively constant at the new levels through the remainder of the 120-day test.

Each of the above effects is fully reversible (unless the polymer is highly filled or plasticized). Once the polymer is removed from contact with the solvent, its physical properties will return to baseline values after a period of time. The 9090 test does not discriminate well between chemical-induced degradation and diffusion-driven swelling. This is because of the requirement that immersed samples be tested within a strict time limit after removal from the leachate bath. If taken at face value, results can be misinterpreted to indicate that true, irreversible chemical degradation of the polymer structure took place.

It might be argued that diffusion-driven swelling improves the physical properties of an HDPE liner:

- The material becomes more flexible and pliable(lower modulus);
- The material may be less susceptible to environmental stress cracking (because of lower internal stresses).

However, the diffusion process is accompanied by transfer of vapor through the barrier by permeation. There will be a gradual buildup of the organic permeant on the reverse side of the geomembrane. The rate or quantity of permeation depends on many factors, including chemical composition of the leachate, concentration gradient and ambient temperature. This is as area of great potential concern to designers and operators of a waste containment facility.

The evaluator should recognize the possible influence of diffusion processes when evaluating data which show an apparent shift in stress-strain dependent properties, especially with highly crystalline materials such as HDPE. Typical "fingerprinting" testing such as thermal analysis will not detect the effects of diffusion, since the high temperatures used will quickly drive off absorbed organic vapors.

Inferences and conclusions not consistent with reported index properties are sometimes made based on "fingerprinting" data. Consider the case in which fingerprinting data show no change, but trends in index properties might suggest that degradation took place. It may be then concluded that the physical test results merely measure the variability of the materials as manufactured and therefore should be given little weight in the analysis. This point of view is short-sighted and does not recognize the limitations of thermal analytical techniques for detecting swelling and softening caused by reversible diffusion.

If diffusion phenomena are suspected, we recommend that further testing be performed to determine whether or not the observed changes are reversible, thus verifying the effects of mutual solubility.

Stress-Strain Measurement and Modulus of Elasticity

Method 9090 requires that for unsupported sheet liner materials modulus of elasticity be measured at each exposure intervals for comparison with baseline values. For semi-crystalline materials, the method requires that ASTM D 882 be used "(Standard Test Methods for Tensile Properties of Thin Plastic Sheeting"). D 882 has .010" (.25 cm) maximum thickness requirement, based on the definition of "film"; therefore strictly interpreted, this standard is not appropriate for application to HDPE liners which are usually .040" or greater in thickness as specified for waste containment facility construction. ASTM D 882 requires a .2" wide (5 mm) strip specimen, with gauge length determined by testing machine gripping points. The alternative method is ASTM D 638, "Standard Test Method for Tensile Properties of Plastics. " The principal feature distinguishing this test from the D882 test is the specimen configuration: D638 requires a dumbbell-shaped specimen.

Careful analysis of stress-strain curves for semi-crystalline HDPE materials indicates that the selection of a straight-line fit which unambiguously defines the elastic modulus is impossible. Figure 1 illustrates a representative stress-elongation curve for a dumbbell tensile specimen cut from HDPE geomembrane sheet material. Figure 2 magnifies the initial portion of that curve as determines using a servo-hydraulic materials testing instrument operating under closed-loop strain control. Strain is measured and strain rate controlled by means of a direct-mounting extensometer with gauge length of one inch (2.54 cm). The curve of Figure 2 demonstrates true visco-elastic behavior as predicted by the classical spring-and dashpot model.The stress-strain curve, magnified to show the initial portion up to the onset of necking, has a characteristic curved shape without a well-defined linear elastic region or yield point. The characteristic shape of the stress-strain curve before yield is very sensitive to leachate-polymer interactions, particularly if the mechanism is diffusion caused by mutual solubility of the polymer with the chemical challenge.

Lab experience has shown that determination of initial stress-strain properties is dependent on measurement technique. In routine testing of HDPE liners, strain is usually measured as a function of crosshead position, and the proportional specimen elongation is inferred from the change in position of the grips. When strain is measured via a direct mounting extensometer, much more accurate strain values can be determined, since factors such as machine compliance, grip slippage, and uncertainties in crosshead position measurement are eliminated. But, stress-strain curves determined through direct extensometry yield even

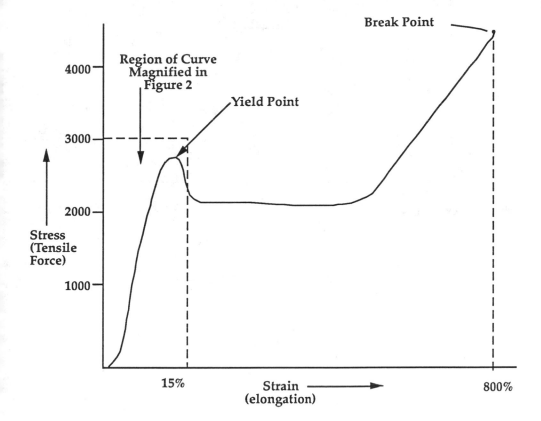

Figure 1:
Typical Stress-Strain Behavior of HDPE Geomembrane

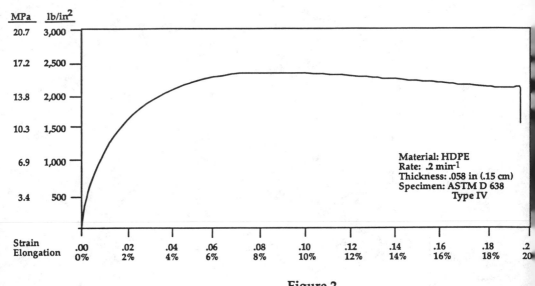

Figure 2
Typical Stress-Strain
Curve for HDPE Geomembrane
Determined by Direct Extensometry

more ambiguous modulus values, since the visco-elastic behavior of the material is much more accurately demonstrated.

Secant modulus has been proposed as a way to normalize modulus values between different laboratories, and to provide a more repeatable index test. This technique is presently under study by ASTM D-35 for application to HDPE liners. If it can be shown that the test gives acceptable inter-laboratory precision, we strongly recommend that the secant modulus calculation technique be substituted for D882 or D638 modulus of elasticity where required by specifications or test methods such as Method 9090 for HDPE materials.

Volatiles and Extractables Determination for HDPE

The Method 9090 test for volatiles and extractables content involves gravimetric analysis which, in the case of HDPE samples, often generates data that vary to the extent that it is anomalous when considered in the context of observed changes for other physical and chemical property data. Typically, the percent volatiles and extractables determined for an HDPE sample is less than the minimum detection limit defined by the test method. This is due to the semi-crystalline polymer structure which allows only small amounts of chemical permeant to be absorbed relative to the weight of the geomembrane specimen.

Quality Assurance in Chemical Compatibility Testing

Experience has shown that formalized quality assurance and quality control systems are extremely critical to the success of a chemical compatibility testing program. We have found that one minor error in transcription can throw results for an entire six-month project into doubt, causing problems for all parties involved. The large quantity of test data produced for a typical chemical compatibility test invites error at any stage from testing and data reduction through reporting and plotting of results. For example, a Method 9090 test of one geomembrane against two leachates at two temperatures will require about 500 individual specimen physical tests and an equal number of reported values, not considering averaging, summarization and plotting. The problem multiplies for geotextiles, since much larger numbers of replicate specimens must be tested. QA/QC procedures may be adapted form those developed for routine analytical testing work and should be consistently applied.

Formal documentation of QA/QC and standard operating procedures for each test method are recommended. Standardization of QC procedures for laboratories performing chemical compatibility testing of geosynthetics would help to increase confidence in decisions based on test results.

Cost

Chemical compatibility testing can be quite expensive if large matrix of materials and leachates are to be tested.

However, there exists a healthy competitive environment and projects are nearly always open to competitive bidding. We hope that the information provided in this article may help to clarify some of the factors driving the cost of testing which must be passed along to the client.

Recommendations

The following recommendations are offered:

1. An alternate method for determination of volatiles and extractables testing should be specified HDPE materials which provides better control of experimental factors that may interfere with the quantification of absorbed volatiles.

2. Modulus of elasticity measured by ASTM D638 or ASTM D882 for HDPE as required by Method 9090 should not be used as a specification minimum value, or applied to evaluate compatibility results. If inter-laboratory agreement can be shown by using secant calculation techniques, this determination should be substituted for the standard modulus of elasticity.

3. Evaluation of chemical compatibility data should not performed using a strict rule set or property loss criterion, since the variability of materials and physical test methods can result in a "fail" decision when no real degradation occurred. Each chemical compatibility test should be evaluated on a basis of all physical and chemical analytical test results considered together. Reasoned scientific judgment, together with statistical analysis of cumulative results, must be applied to confirm that the materials tested were or were not affected by the exposure.

REFERENCES

[1] EPA Method 9090, "Compatibility Tests for Wastes and Membrane Liners" in EPA SW-846, Test Methods for Evaluating Solid Waste, U.S. Environmental Protection Agency, Washington, D.C.

Bruce E. Dudzik and Louis G. Tisinger

AN EVALUATION OF CHEMICAL COMPATIBILITY TEST RESULTS OF HIGH DENSITY
POLYETHYLENE GEOMEMBRANE EXPOSED TO INDUSTRIAL WASTE LEACHATE

REFERENCE: Dudzik, B.E. and Tisinger, L.G., **"An Evaluation of Chemical Compatibility Test Results of High Density Polyethylene Geomembrane Exposed to Industrial Waste Leachate,"**Geosynthetic Testing for Waste Containment Applications, ASTM STP 1081, Robert M. Koerner, editor, American Society for Testing and Materials, Philadelphia, 1990.

ABSTRACT: HDPE geomembrane was exposed to leachate generated from an existing hazardous waste landfill for up to one year both in the laboratory (EPA Method 9090) and in a leachate collection sump at the site. Various mechanical, physical, and microstructural properties were monitored after 30, 60, 90, 120, and 365 days of exposure in the laboratory and after 120 and 365 days in the sump. Comparison of the tensile property results among the laboratory and sump-exposed samples showed considerable similarity. Variability in the geomembranes tested was evident. Test results are presented, along with a discussion on measuring the apparent variability of the geomembrane.

KEYWORDS: geomembranes, compatibility testing, EPA Method 9090, field testing, microstructural properties.

Part 264.301(a)(1)(i) of Title 40 of the Code of Federal Regulations requires that chemical compatibility testing of the geomembranes used as liners in hazardous waste landfills must be performed to demonstrate that the geomembrane is suitable for use in the primary or secondary liners. Chemical compatibility testing consists of exposing the geomembrane to the leachate generated from the landfill and subjecting the exposed samples to physical/mechanical property tests. The results of these tests are then compared to the data obtained on unexposed (to leachate) samples to determine if exposure to the leachate has significantly affected the physical/mechanical properties of the material.

Laboratory compatibility testing was performed, in accordance with EPA Method 9090[1], on three geomembrane samples exposed to leachate for immersion periods of 30, 60, 90, 120, and 365 days. The following modifications (or deviations) to EPA Method 9090 were made in the testing program described in this paper:

1. Whenever a difference existed between the ASTM test method and Method 9090 regarding the number of specimens to be tested, the ASTM test method was always used.

2. As noted above, the exposure periods for the testing program included immersion testing for 365 days.

In addition, the same three geomembranes were also exposed to leachate in the field by installing perforated stainless steel baskets containing the geomembrane samples directly into

Mr. Dudzik is a senior project engineer at RMT, Inc., Madison, WI 53708; and Mr. Tisinger is Program Manager, Chemistry at GeoSyntec, Inc. of Boynton Beach, FL 33426.

one of the leachate collection sumps at the existing hazardous waste landfill. The field samples were exposed to leachate for 120 and 365 days to correspond to the last two immersion periods for the Method 9090 testing. The same tests were performed on the field samples as those for laboratory immersion samples. A comprehensive bibliography on compatibility testing procedures, geomembrane testing procedures, and results of previous testing programs can be found at the end of each section in the Matrecon, Inc.[2] report.

The three polyethylene geomembranes tested, along with an identification of the resin type used in manufacturing, are as follows:

Designation	Thickness-mm	Resin Type
1	2.0	High-Density
2	2.5	High-Density
3	2.0	Linear Medium-Density

All three of the tested geomembranes contain 2-3 percent by weight of carbon black in addition to other proprietary antioxidants and fillers. Samples were provided by the manufacturers and shipped directly to the laboratory for testing. Unexposed geomembrane samples were tested to establish a baseline for comparison of the test results from leachate-exposed samples. All immersion cells used for the laboratory testing were mechanically stirred to prevent concentration gradients of any leachate constituent from building up at or near the surface of the geomembrane samples.

The purpose of this paper is to 1) provide a description of the methodology used for containing and installing the field samples; 2) compare the test results obtained for the tensile properties of the field and laboratory samples exposed to leachate; and 3) provide an analysis of the test results that allows for a distinction to be made between material degradation and variability.

FIELD SAMPLES

The field sample testing program was designed to supplement the laboratory compatibility testing done in accordance with using EPA Method 9090. Laboratory testing, while more controllable than field testing, poses several problems as discussed below:

1. Contact of samples with fresh leachate is not possible with laboratory testing. Method 9090 states that leachate replacement can be used to reduce this concern, but practical considerations resulted in Method 9090 adopting a once-every-30-day interval for leachate replacement.

2. Volatile organic compounds are lost due to leachate handling (including leachate replacement at 30-day intervals), transportation, and setup in the immersion cells. Additional volatile compound losses will occur due to the fact that it is impossible to provide a perfect seal between the cell top and the mechanical stirring shaft used in the immersion cells.

3. Method 9090 is predicated on the assumption that individual constituents and their concentrations in leachate do not vary with respect to time. Actual landfill leachate data, collected over time, suggest that the constituents and their concentrations are always changing. By conducting field testing, it is therefore possible to have the geomembranes exposed to a much greater range of organic and inorganic constituents.

Field testing of geosynthetics eliminates the above problems by exposing the samples to flowing leachate, as it is actually generated at the site, provided that satisfactory means can be made available to locate the field samples. Field samples represent a "real world" scenario that is virtually impossible to duplicate in the laboratory. The same physical and microstructural testing program was performed on the field samples as that used for Method 9090 testing.

In their draft manual on methodology for removable field samples, Arthur D. Little[3] points out that 1) samples should be mounted in such a way so as to prevent any sample from touching its neighbor; and 2) only samples of the same material, by manufacturer, should be placed in any one rack.

The original scope of services for field testing adhered to the above criteria with the result that 26 perforated stainless steel baskets, each 20 inches long by 11 inches wide by 7 inches deep, would have been required for installing both the 120-day and 1-year field sample sets. Installation of 26 baskets at the landfill site would have required a modification to the operating permit in order to provide a minimum water depth of 24 inches in the leachate collection sump to provide for complete immersion of the baskets in the sump.

The request to modify the operating permit for the landfill by increasing the allowable level in the leachate collection sump was denied by the permit-issuing authority. The denial forced a modification to the field sampling program that reduced the number of baskets required to a total of 10 for both the 120- and 365-day periods. The revised plan differed from the Arthur D. Little report in that 1) samples could touch one another as long as the samples were of the same material (the overall testing program included additional geosynthetics); 2) samples of different geosynthetics could be placed in the same rack; and 3) samples did not have to be hung with mounting hooks, but rather the samples could be placed on top of one another. Inclusion of these changes into the sump testing program were not expected to have any effect on test results.

Each set of field samples (120-day and 1-year) consisted of five baskets, each of which were constructed of 316 stainless steel with a bolted stainless steel cover. To simplify installation of the baskets into the sump, the individual cover bolts for each rack were removed, and all five racks were mounted together at each corner using 30-inch lengths of 1/8-inch threaded stainless steel rod. The complete assembly is shown on Figure 1.

After securing the baskets into one assembly, 1/4-inch stainless steel cable and 3/8-inch polypropylene rope were attached to opposite corners of the assembly. The two assemblies were lowered into the sump with the ends of both the cable and rope tagged for each sample. The two assemblies were placed side by side in the sump. The ends of the cables and ropes were tied to the underside of the sump cover. Both sets of samples were installed on January 22, 1988.

The 120-day and 365-day field samples were removed from the sump on May 23, 1988, after 121 days of exposure, and on January 23, 1989, after 365 days of exposure. The following procedure was used to remove the racks from the sump and ship them for testing:

1. The rack assembly was brought to the surface, while remaining in the sump, and washed with tap water. After rinsing with water, each assembly was allowed to drain for 20-30 minutes.
2. Each assembly was pulled completely from the sump, placed on a plastic sheet next to the sump, and visually inspected. There was no visual evidence of damage; however, the stainless steel racks were visibly discolored dark black over most of their surface.
3. The assembly was packed, without disassembling the individual racks, in five large polyethylene bags. The assembly was then placed in a cardboard box, taped, and then shipped overnight for immediate testing.

LEACHATE CHARACTERIZATION

Table 1 lists the general inorganic and organic constituents found in the leachate used for the Method 9090 testing program. The leachate used for this compatibility program falls within the overall range of leachate characteristics presented in Section 2.2.4 of the Matrecon, Inc.[1], report with the exception that leachate analysis for organic constituents was performed by GC/MS using EPA Methods 624 for volatile organics and Method 625 for base neutral and acid extractables. Up to 10 additional major peaks for Method 624 or 625, for compounds with molecular weights up to 750, were identified in addition to those constituents normally included in Methods 624 and 625.

TESTS AND TEST METHODS

Table 2 lists the tests and test methods that were performed on each of the three geomembranes. In cases where EPA Method 9090 and the referenced ASTM test method differed regarding the number of specimens to be tested, the ASTM testing procedure was used. Two general types of testing were included in the compatibility testing program.

The first type is the physical/mechanical property testing which is required by EPA Method 9090. This includes, among others, tensile properties, puncture resistance, and tear resistance. The results of mechanical property tests in compatibility studies have been typically interpretated on the assumption that any changes in properties from unexposed to exposed specimens are caused by degradation of the polymers used to fabricate the geomembranes.

The second type of testing falls into the category of microstructural analysis, and includes tests such as crystallinity and oxidative induction temperature (by differential scanning calorime-

FIG. 1-*Perforated stainless steel basket assemblies used for holding field samples prior to installation in leachate collection sump.*

try), percent composition (by thermal gravimetric analysis), and material structure (by infrared spectroscopy). These tests evaluate, on a microscopic level, the chemical composition of the material, the structure of the polymer, and the effectiveness of additives such as antioxidants. This type of testing examines the condition of the polymer itself rather than measuring the physical and mechanical changes of the bulk material.

The combination of both types of testing, therefore, allows for the determination of whether changes in physical and mechanical properties are the result of actual leachate attack or degradation, or the result of variations in the product through the various steps employed in the manufacturing process.

Table 1 -- Leachate characteristics[a]

Parameter	Units	Value
Alkalinity, Total	mg/l as $CaCO_3$	11600
BOD_5	mg/l	19500
COD	mg/l	37200
Conductivity	μMhos	38575
Oil & Grease	mg/l	210
pH	St. Units	9.1
Silica	mg/l as SiO_2	141
Total Dissolved Solids	mg/l	50,100
Total Organic Carbon	mg/l	11500
Total Suspended Solids	mg/l	212
Turbidity	NTU	156
Calcium	mg/l	30
Iron	mg/l	19
Magnesium	mg/l	16
Nickel	mg/l	28
Potassium	mg/l	2715
Sodium	mg/l	13250
Chloride	mg/l	12100
Nitrate	mg/l as N	51
Phosphorus	mg/l as P	21
Sulfate	mg/l as SO_4	3850
Benzene	μg/l	6500
Chloroform	μg/l	1330
1,1 Dichloroethane	μg/l	2900
Ethyl Benzene	μg/l	35
Phenol	μg/l	14100
Styrene	μg/l	95
Toluene	μg/l	21100
m-Xylene	μg/l	284
o-Xylene	μg/l	93

[a] Results for volatile organics (EPA Method 624) and Base/Neutral Organics (EPA Method 625) are shown only for those constituents with concentrations above their respective detection limit.

TABLE 2 -- Tests For geomembranes

Test	Defined Procedure Number[a]
Carbon Black Content	ASTM D1603
Carbon Black Dispersion	ASTM D3015
Crystallinity	ASTM E793
Environmental Stress Cracking	Modified ASTM D1693[b]
Extractables	ASTM D3421
Gauge Thickness	ASTM D374 Method C
Hardness	ASTM D2240 Shore Durometer Type D
Infrared Spectroscopy[c]	
Length	Section 7.6 of Method 9090
Mass (Water Absorption)	Section 7.6 of Method 9090
Melt Index	ASTM D1238
Modulus of Elasticity	ASTM D638 Type IV Specimen[d]
Mullen Burst	ASTM D3786
Puncture Resistance	FTMS 101C Method 2065
Shear and Peel Seam Strength	ASTM D413 and ASTM D882
Specific Gravity	ASTM D792
Tear Resistance	ASTM D1004
Tensile Properties	ASTM D638, Type VI Specimen[e]
Thermal Gravimetric Analysis[f]	
Volatiles	ASTM D2288
Water Vapor Transmission	ASTM E96 Method BW
Width	Section 7.6 of Method 9090

[a]Defined procedure numbers refer to the appropriate section of EPA Method 9090, entitled "Compatibility Test for Wastes and Membrane Liners."

[b]Entire test conducted in leachate at 30-, 60-, 90-, 120-, and 365-day exposures. Surfactant (Igepal), referenced in procedure, was not used.

[c]Infrared spectroscopy was performed on a thin-layer microtome of the sample.

[d]Immersion periods will be for 30, 60, 90, 120, and 365 days, and are to be performed in accordance with EPA Method 9090.

[e]An extension rate of 2 inches per minute was used, and the modulus test will be computer-calculated (at steepest part of the stress-strain curve).

[f]Charles M. Earnest, "Modern Thermogravimetry," Analytical Chemistry, 56 (13):1471A, 1984.

The remainder of this section describes in greater detail the applicability or methods used for differential scanning calorimetry, infrared spectroscopy, and tensile properties. Carbon black content and dispersion and melt index were performed on unexposed samples only as part of the conformance testing program. Infrared spectroscopy was performed on unexposed and the 120- and 365-day laboratory and field samples; all other tests were performed at each time interval for field and laboratory samples.

Differential Scanning Calorimetry

Differential scanning calorimetry (DSC), for determining crystallinity, measures the amount of heat required to maintain a test specimen at the same temperature as a reference specimen heated at a constant rate of increasing temperature. Two types of tests are typically performed with DSC on geomembranes. Degree of crystallinity is determined from the endotherms, and the oxidative induction temperature (OIT) is determined from the exotherms.

Degree of crystallinity is a measure of the molecular orientation of a material, relative to a perfectly crystalline reference substance (Indium). The test actually determines heat of fusion of the specimen, which is then divided by the heat of fusion of Indium to obtain the degree of crystallinity expressed in percent. All sampling was conducted from the surface of the geomembrane specimens to ensure that the superficial extent of the leachate exposure would be observed.

Polyethylene crystallinity is directly proportional to the density or specific gravity. It is also worth noting that the variability associated with the mechanical properties testing is closely related to the variability of the crystallinity, since many of the mechanical properties are related to material structure.

Oxidative induction temperature (OIT) is a measure of the temperature at which reaction of the specimen with oxygen occurs. The temperature recorded is the temperature corresponding to the onset of the exotherm (the onset being the point of initial change in slope). This test measures the effectiveness of antioxidants added to the PE resin in the geomembrane manufacturing process. Most PE materials typically contain less than 2 percent of antioxidants. It is also important to note that, while both antioxidants and carbon black are present to reduce ultraviolet attack, they accomplish their respective jobs by totally different means, since carbon black is an ultraviolet absorber.

Infrared Spectroscopy

Infrared (IR) spectroscopy analysis allows for a determination of the structural characteristics of the polymer. IR is employed to identify new spectral bands that are consistent with a particular degradative process and that are not characteristic of the base material. This test was included to assist in the Method 9090 testing to differentiate whether widely variable mechanical properties are due to material degradation by leachate exposure (where IR analysis would show new bands) or to variability due to the manufacturing process (where no new bands would be present). This test can also be used to identify any constituents from leachate exposure which may become absorbed into the geomembrane polymer matrix.

Tensile Properties

The compatibility testing conducted in this study included determination of the tensile properties of the polyethylene (PE) geomembrane materials. The following tests were performed:

1. Yield stress
2. Yield elongation
3. Break stress
4. Break elongation
5. Modulus of elasticity

All of the above tests were performed using ASTM D638 with Type IV specimens. Practice within the geomembrane industry has varied as to the method of calculating several of the above tests. The calculations for yield and break stresses are based on specimen width and thickness, neither of which is subject to interpretation within the industry. The calculation for break elongation is based on the specimen length, which is currently subject to different interpretations within the industry. The different interpretations concern whether the calculations are based on the "length of narrow section" or the "distance between grips." (Both terms are defined by Figure 1 of ASTM D638 for Type IV specimens.) For Type IV specimens, the length of narrow section is 1.3 inches, and the distance between grips is 2.5 inches. (Yield-elongation calculations are all performed using the 1.3-inch length of narrow section dimension). The calculated break elongations are based on a specimen length of 1.3 inches. To convert these values to test results based on specimen lengths of 2.5 inches, the test value should be multiplied by 0.52; to convert to test specimens of 2 inches, the values should be multiplied by 0.65.

The determination of a specimen's modulus of elasticity requires that the slope of the initial part of the stress-strain diagram be calculated. Since PE-based materials do not have an initial linear stress-strain curve, the determination of this slope is somewhat subjective. Modulus of elasticity testing was performed by a computer-calculated system in which the computer calculates the slope at the steepest part of the initial curve and uses that value in calculating the specimen's modulus of elasticity.

TEST RESULTS

Three sets of results for the compatibility testing program are presented in this section covering 1) microstructural testing which can be used to determine if the polyethylene was degraded by exposure to leachate; 2) a comparison of the results obtained for the 120- and 365-day field and laboratory sample tensile properties; and 3) a comparison of results obtained on nondestructive and destructive tests which are used to determine the importance of manufacturing process variability when assessing the results of chemical compatibility testing.

The purpose of performing the compatibility tests is to determine if the geomembrane is affected by exposure to leachate over time. This determination is summarized in this section by first reviewing those test results that indicate if degradation of the PE has occurred, and then by comparing the results of nondestructive and destructive tests.

Degradation of the PE would be indicated by the following:

1. The results of the OIT and IR tests would show whether the geomembrane was being oxidized by the leachate.
2. The results of the IR would also indicate if constituents from the leachate were being absorbed into the PE matrix.

Differential Scanning Calorimetry

Two tests were performed using differential scanning calorimetry (DSC) -- crystallinity and oxidative induction temperature. Figure 2 graphs the crystallinity data for each of the geomembranes throughout the 365-day Method 9090 tests, including data for both the 23° and 50°C immersion temperatures. The crystallinity is a function of the density of the base resin; therefore, it should be slightly higher for Materials #1 and #2 (manufactured from HDPE) than for Material #3 (manufactured from LMDPE). There is a relatively significant degree of variability for the crystallinity data; however, no time trends in the data are evident for the graph.

Crystallinity data (expressed in percent by weight) for the 120-day and 365-day Method 9090 and field samples are listed below in Table 3.

TABLE 3 -- Field and laboratory sample crystallinity test results

Time & Sample Name	Material #1	Material #2	Material #3
Unexposed	52.75	37.86	63.88
120-Day 9090-23°C	52.98	48.13	56.64
120-Day 9090-50°C	52.81	40.72	28.27
120-Day Field Sample	49.68	51.84	34.03
365-Day 9090-23°C	40.08	50.79	37.16
365-Day 9090-50°C	43.44	38.64	39.90
365-Day Field Sample	39.43	40.73	40.61

The above results show large variation between Materials #1 and #2, with the 80-mil thickness having the higher crystallinity. In addition, the crystallinity of Material #3 is less than that for the geomembranes manufactured from HDPE, as expected, except for the unexposed and 120-day Method 9090-23°C results which are higher than any of the other values for materials #1 and #2. Material #3 shows the greatest variability within materials. There are no apparent time trends in the above data for any of the geomembranes tested.

Oxidative induction temperature (OIT) data are presented on Figure 3 for the Method 9090 testing in 23°C and 50°C immersion cells. The OIT data are quite consistent throughout the

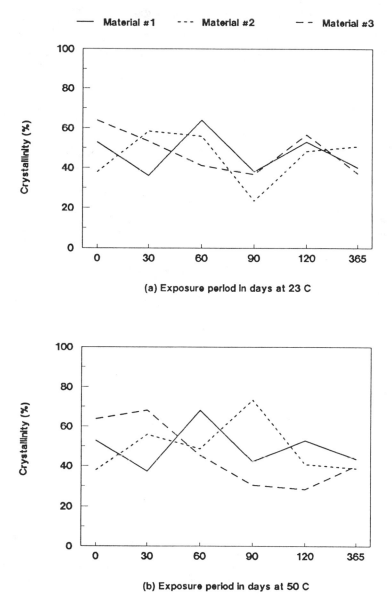

FIG. 2-*Crystallinity Test Results.*

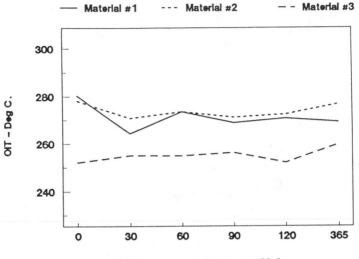

(a) Exposure period in days at 23 C.

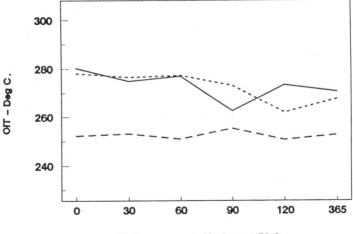

(b) Exposure period in days at 50 C.

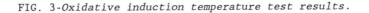

FIG. 3-*Oxidative induction temperature test results.*

immersion periods for each geomembrane and are also quite consistent among geomembranes; the overall range of OIT data on the geomembranes was from 250.6 to 277.3°C. OIT data (expressed in °C) for the 120-day and 365-day Method 9090 and field samples are listed below in Table 4.

The data from Table 4 do not indicate any significant changes or time trends in the values for OIT for the three tested geomembranes. Decreased antioxidants would be indicative of either a strong oxidizing agent in the leachate that would, over time, deplete the phenolic-based antioxidants or migration of the antioxidants from the geomembranes. This reaction is not the same as that for ultraviolet radiation attack of polyethylenes. Individual OIT curves are also flat, indicating little change in antioxidants and no apparent degradation of any of the geomembranes tested. Compared with the two HDPE geomembranes, the lower OIT of Material #3 is typical

TABLE 4 -- Field and laboratory sample OIT results

Time & Sample Name	Material #1	Material #2	Material #3
Unexposed	280.3	278.1	252.2
120-Day 9090-23°C	270.7	272.4	252.0
120-Day 9090-50°C	273.4	262.0	250.6
120-Day Field Sample	266.6	278.6	262.9
365-Day 9090-23°C	269.2	276.7	259.6
365-Day 9090-50°C	270.6	267.6	252.6
365-Day Field Sample	275.2	278.0	261.0

of comparisons made between products using LMDPE and HDPE. The above results do not indicate oxidation of the geomembranes as a result of leachate exposure.

Infrared Spectroscopy

One of the principal degradative processes, indicated by infrared spectroscopy (IR), would be oxidation of the polymer; however, this would also be indicated by correspondingly lower OIT values, as discussed above. Oxidation of the PE would be indicated by new bands within the 1600 to 1800 cm^{-1} wavenumber region in the IR spectrum. No new bands are present on any of the 120- or 365-day sample IR scans, including the 1,600 to 1,800 cm^{-1} band or other bands indicating absorption of any organic components. The IR results indicate that organic constituents of the leachate have not been absorbed into the geomembrane polymer matrices to any significant extent. The fact that the IR and OIT tests indicate no degradation allows for the conclusion that physical property variations discussed elsewhere are caused by variations in the manufacturing processes of these materials. The size and number of IR spectra preclude their inclusion in this paper.

Tensile Property Results on Field and Laboratory Samples

Tables 5 through 7 summarize the results, by material, of tensile property testing performed on both laboratory and field samples at the 120- and 365-day leachate exposure intervals.

Destructive and Nondestructive Test Results

Table 8 presents the results (expressed as percent change for each time interval compared to the unexposed sample value) of most of the geomembrane strength tests (all of which are destructive), and Table 9 presents the results of the nondestructive tests. The results

Table 5 -- Comparison of tensile property results on laboratory and field samples for Material #1

Property	Units	Method 9090 120 Days-23°C	Method 9090 120 Days-50°C	120 Day Field	Method 9090 365 Days-23°C	Method 9090 365 Days-50°C	365 Day Field
Break Stress	MPa						
Machine Direction		26.9	23.9	25.7	28.2	26.3	26.2
Cross Direction		29.3	24.6	24.7	24.7	21.2	27.0
Yield Stress	MPa						
Machine Direction		18.9	19.7	18.5	17.7	20.6	18.4
Cross Direction		19.1	22.3	20.1	18.2	20.7	18.8
Break Elongation	%						
Machine Direction		1125	1054	1113	1223	1168	1125
Cross Direction		1244	1088	1086	1106	1040	1210
Yield Elongation	%						
Machine Direction		15.0	15.7	15.4	15.7	17.1	15.8
Cross Direction		14.9	15.2	14.3	14.8	16.6	15.3
Mod. of Elasticity	MPa						
Machine Direction		480	497	563	503	562	513
Cross Direction		478	458	561	503	510	562

Table 6 -- Comparison of tensile property results on laboratory and field samples for Material #2

Property	Units	Method 9090 120 Days-23°C	Method 9090 120 Days-50°C	120 Day Field	Method 9090 365 Days-23°C	Method 9090 365 Days-50°C	365 Day Field
Break Stress	MPa						
Machine Direction		27.4	25.2	25.6	29.5	21.3	30.0
Cross Direction		28.9	22.7	28.0	27.3	15.9	30.7
Yield Stress	MPa						
Machine Direction		18.1	19.4	19.0	17.4	19.9	17.7
Cross Direction		18.0	19.0	18.2	17.8	20.5	17.9
Break Elongation	%						
Machine Direction		1208	741	1147	1314	1013	1319
Cross Direction		1288	1080	1265	1263	854	1360
Yield Elongation	%						
Machine Direction		16.7	17.2	17.4	18.1	17.8	16.9
Cross Direction		16.0	15.9	17.0	17.8	17.4	17.1
Mod. of Elasticity	MPa						
Machine Direction		413	466	598	523	669	510
Cross Direction		406	462	581	563	590	503

Table 7 -- Comparison of tensile property results on laboratory and field samples for Material #3

Property	Units	Method 9090 120 Days-23°C	Method 9090 120 Days-50°C	120 Day Field	Method 9090 365 Days-23°C	Method 9090 365 Days-50°C	365 Day Field
Break Stress	MPa						
Machine Direction		26.6	24.6	25.3	25.4	23.7	25.5
Cross Direction		19.4	13.8	24.5	25.6	17.0	25.6
Yield Stress	MPa						
Machine Direction		18.5	19.1	17.2	17.8	17.8	17.1
Cross Direction		19.2	19.4	17.8	17.2	19.0	17.5
Break Elongation	%						
Machine Direction		1229	1160	1234	1234	1157	1229
Cross Direction		1071	863	1243	1317	970	1421
Yield Elongation	%						
Machine Direction		14.9	16.2	15.7	15.0	16.6	16.0
Cross Direction		14.0	15.4	14.6	15.1	15.9	15.2
Mod. of Elasticity	MPa						
Machine Direction		474	439	560	572	605	456
Cross Direction		474	424	581	581	617	435

Table 8 -- Comparison of results for geomembrane strength tests[a]

Exposure Period	Hardness	Break[b] Stress	Elongation at Break[b]	Mullen Burst
Material #1				
23°C Immersion Temperature				
30-day	+ 2.3	-10.8	- 6.9	- 0.6
60-day	+ 1.0	- 0.3	+ 6.1	-21.9
90-day	+12.0	-18.5	- 7.6	+ 1.8
120-day	+15.9	-14.5	- 6.7	+ 4.1
365-day	+ 2.6	-10.5	+ 1.5	- 4.4
50°C Immersion Temperature				
30-day	+ 1.0	-15.8	-15.5	- 4.7
60-day	+ 1.0	- 1.2	+ 0.5	-10.8
90-day	+ 8.4	-42.6	-19.8	+16.1
120-day	+11.0	-24.2	-12.6	+33.9
365-day	+ 2.6	-16.5	- 3.1	0.0
Material #2				
23°C Immersion Temperature				
30-day	- 3.1	+17.3	+14.8	- 2.3
60-day	- 2.2	+30.4	+ 8.9	-22.8
90-day	+ 8.4	+18.7	+20.4	- 1.8
120-day	+ 5.3	+19.9	+22.2	- 0.6
365-day	+ 0.9	+29.3	+32.9	+ 4.1
50°C Immersion Temperature				
30-day	- 1.9	+ 4.7	+ 1.3	+ 5.2
60-day	- 3.1	+32.8	+18.6	-22.8
90-day	+10.0	-16.4	- 0.5	+ 8.6
120-day	+ 8.7	+10.6	-25.0	- 3.8
365-day	- 0.3	- 6.4	+ 2.4	+12.4
Material #3				
23°C Immersion Temperature				
30-day	+ 0.3	- 8.8	- 4.7	- 1.2
60-day	+ 1.9	- 1.7	+ 4.2	-35.7
90-day	+19.4	-30.4	-19.5	- 3.3
120-day	+14.5	- 8.1	- 4.6	+ 2.8
365-day	+ 2.3	-12.2	- 4.2	- 3.1
50°C Immersion Temperature				
30-day	+ 0.6	-16.3	-11.6	-31.0
60-day	- 0.6	-18.3	-11.6	-31.0
90-day	+13.9	-42.9	-25.8	+ 3.1
120-day	+14.8	-15.0	- 9.9	+ 4.1
365-day	+ 2.6	-18.0	-10.2	- 1.3

[a] Percent change compared with unexposed sample
[b] Based on data for machine direction.

presented in Table 9 would show pure variation due to degradative processes attacking the PE, whereas results in Table 8 are influenced by both degradative changes and variability associated with the manufacturing processes. None of the data presented in either table show any time trends during the 365-day testing program. (Any time trends in Table 9 would be a clear indication of changes resulting from degradative attack.) The most striking difference in comparing Tables 3 and 4 is the magnitude of the percent changes shown. With the exception of material thickness, the percent changes range from -2.4 to +1.0. Compared to the other tests on Table 9, thickness data show the greatest variation ranging from -4.4 to +13.1 percent. The values on Table 8, however, range from -42.9 to +33.9 percent, thus indicating a much greater degree of variability than the test results shown on Table 9.

CONCLUSIONS

The results of the IR and OIT tests indicate that no degradation of any of the three geomembranes by either oxidation or absorption of constituents occurred. Furthermore, absorption of chemicals into the polymer matrix would have a plasticizing effect on the geomembranes. Plasticization would be indicated by decreasing trends in the crystallinity, hardness, and tensile strengths at yield and break, and by increasing trends in specific gravity and elongation at yield and break. Actual analysis of these test results shows that there are no time-dependent trends for any of the tests.

Another degradation mechanism of PE is cross linking. Cross linking can be described as additional ties forming between layers of the PE polymer structure. Cross linking would be indicated by significant decreases in the crystallinity accompanied by significant increases in the hardness and specific gravity. With the possible exception of an increasing hardness trend in the 0- to 120-day data, none of these parameters exhibited a significant time-trend behavior over the 0- to 365-day period. Analyses of the specific gravity and crystallinity data shows no time trends, thus ruling out cross linking over this period.

OIT, IR, and the combinations of tests for cross linking and plasticization address most types of degradative processes; the results, therefore, of this testing program showed that material degradation was not evident and that the changes noted in physical/mechanical/microstructural properties were due to differences in manufacturing variability.

The comparison of results obtained on field and laboratory tensile property tests shown on Tables 5 through 7 can be summarized as follows:

1. The results obtained on field samples, in general, were not as variable as those obtained from the Method 9090 laboratory testing.
2. Field sample values, for each time interval, in general, agreed well with those obtained for the 23°C laboratory test, and both of these results were generally higher than those measured for the 50°C Method 9090 immersion testing. The overall magnitude of the differences between 23° and 50° data are, for the most part, below the standard errors for the respective test.
3. The overall agreement between laboratory and field samples indicates that, for this study, Method 9090 was able to simulate field conditions in a laboratory test. In addition, the results for this study show that little difference probably exists in testing performed using a single leachate in a static environment (Method 9090) versus leachate exposure resulting from a continuously changing leachate (field testing).

Overall, the data from Tables 5 through 7 indicate the same magnitude of manufacturing process variability as those indicated on Table 8, thus indicating that the 120- and 365-day data were not substantially different than the test results obtained at the 30-, 60-, and 90-day intervals in the Method 9090 testing.

EPA's FLEX[4] system for evaluating chemical compatibility testing of geomembranes does not presently recognize material variability as a component in analyzing test results. At the present time, FLEX assumes that all changes in material properties are due to degradation of the material from leachate exposure. The results presented in this paper show that changes in

Table 9 -- Comparison of results for geomembrane
nondestructive tests[a]

Exposure Period	Length	Mass	Thickness	Width
Material #1				
23°C Immersion Temperature				
30-day	0.0	+ 0.2	- 1.2	0.0
60-day	- 0.3	- 0.1	+ 1.2	- 0.8
90-day	0.0	+ 0.2	+13.1	- 0.2
120-day	+ 0.2	0.0	+ 1.2	- 1.1
365-day	- 0.5	- 2.0	- 0.3	- 1.0
50°C Immersion Temperature				
30-day	- 0.8	0.0	- 1.2	- 0.9
60-day	- 0.3	+ 1.1	0.0	- 0.2
90-day	- 0.1	0.0	+ 2.3	- 0.9
120-day	0.0	0.0	+ 1.2	- 0.8
365-day	0.0	0.0	- 0.3	- 0.7
Material #2				
23°C Immersion Temperature				
30-day	0.0	0.0	- 2.0	- 0.7
60-day	0.0	- 0.1	0.0	0.0
90-day	- 0.1	0.0	0.0	- 1.2
120-day	0.0	0.0	+ 3.0	- 1.3
365-day	0.1	0.0	- 0.9	- 1.0
50°C Immersion Temperature				
30-day	0.0	+ 0.1	+ 3.1	0.0
60-day	0.0	0.0	0.0	- 0.1
90-day	- 0.1	+ 0.1	+ 2.0	- 1.2
120-day	0.0	- 1.0	- 1.0	+ 1.0
365-day	0.2	0.0	+ 4.7	+ 0.2
Material #3				
23°C Immersion Temperature				
30-day	0.0	+ 0.4	- 2.4	- 1.3
60-day	+ 0.2	0.0	0.0	- 1.0
90-day	- 0.2	0.0	- 3.5	- 1.3
120-day	+ 0.1	0.0	+ 1.2	+ 0.5
365-day	+ 0.2	0.1	- 2.6	- 0.2
50°C Immersion Temperature				
30-day	- 0.2	+ 0.3	- 3.6	- 2.4
60-day	- 0.1	0.0	- 0.1	- 0.1
90-day	0.0	0.0	- 1.2	- 0.9
120-day	- 0.3	0.0	- 2.3	+ 0.2
365-day	0.0	0.1	- 4.4	- 0.8

[a] Percent change compared with unexposed sample

physical/mechanical/microstructural properties do exist and that these changes are not due to degradation.

REFERENCES

[1]   Matrecon, Inc., *Lining of Waste Containment and Other Impoundment Facilities*, Office of Research and Development, Risk Reduction Engineering Laboratory, EPA Report Number EPA/2-88-052, September 1988.
[2]   Environmental Protection Agency, *Method 9090 - Compatibility Test For Wastes and Membrane Liners*, Technical Resource Document SW-846 Test Methods for Evaluating Solid Wastes, Third Edition 1986.
[3]   Arthur D. Little, Inc., *Proposed Methodology for Removable Coupon Testing*, Draft Final Report Prepared Under EPA Contract Number 68-02-3968, September, 1985.
[4]   Arthur D. Little, Inc., *FLEX Program Manual and User's Guide*, Draft EPA Report, August 1988.

Test Methods and Procedures to Evaluate Geomembranes

Robert L. Gray

ACCELERATED TESTING METHODS FOR EVALUATING POLYOLEFIN
STABILITY

REFERENCE: Gray, R.L., "Accelerated Testing Methods
for Evaluating Polyolefins Stability", Geosynthetic
Testing for Waste Containment Applications, ASTM STP
1081, Robert M. Koerner, editor, American Society for
Testing and Materials, Philadelphia, 1990.

ABSTRACT: Most polymeric materials are subject to
oxidation; the rate of oxidation depends on the
polymer, fabrication methods, and end use
conditions. Oxidation of polyolefins can result
in discoloration and loss of physical properties
such as elongation, impact strength, tensile
strength and flexibility. Antioxidants and light
stabilizers protect polyolefins against oxidation
by controlling molecular weight change and the
ensuing loss of physical and mechanical
properties.

Careful evaluation of the long-term performance of
polyolefins is critical to the successful
development of new stabilization systems. While
real-time outdoor exposure of materials is
desirable, it is usually not practical due to the
exposure times required. Consequently, several
accelerated aging methods have been developed.
This paper will discuss and compare commonly used
accelerated test methods.

KEYWORDS: geosynthetic, aging, durability,
oxidative induction time, long term heat aging

High Density Polyethylene (HDPE) has become widely used in
the United States as the primary material for geomembrane
liners [1,2]. This is due in part to its excellent
chemical resistance [3,4]. Advances in stabilization of

Dr. Gray is a research and development chemist at CIBA-
GEIGY Corporation, 444 Saw Mill River Road, Ardsley, New
York 10502.

polyolefins has made these materials quite durable, however as the required service lives of these liners are often in excess of thirty years, extensive evaluations must be carried out in order to insure the integrity of the liners.

Proper selection of test methods is critical in successfully estimating the anticipated service life of a particular sample.

The durability of geosynthetics is effected by the condition of the material after processing, thermal exposure, light (UV) exposure, and chemical exposure. A variety of tests used to evaluate the physical properties of a liner after processing are well established [5]. The most reliable long term durability testing method would be actual exposure under realistic conditions. Obviously this is not feasible due to the length time that would be required to obtain useful results. Therefore it becomes necessary to develop accelerated test methods which will provide data that can be used to realistically evaluate differing stabilization systems in a timely manner.

Two of the most common accelerated aging methods for evaluating resistance to thermooxidative degradation are oven aging [6] and thermal analysis [7,8]. These methods will be discussed and compared in this paper.

Oven Aging: Long term thermal aging is often carried out in forced air ovens where samples are exposed to relatively high temperatures (temperatures do not however, exceed the melting point of the sample). In this method samples are mounted in ovens and exposed at the appropriate temperature. Replicate samples are periodically pulled from the oven and can be evaluated by a number of methods. Selection of the proper testing temperature is important as temperature can effect the relative performance of stabilizers systems. Evaluation by this method is certainly more time consuming than oxidative induction time (OIT) but as this paper will demonstrate, the results may provide a more reliable insight into expected "real life" field durability.

Oxidative Induction Time: An accelerated test method commonly used in the geosynthetic industry is oxidative induction time. An OIT is determined by measuring the length of time between the introduction of oxygen to a Differential Scanning Calorimeter (DSC) and the onset of the exotherm resulting from the initiation of rapid thermooxidative degradation. It is often assumed that this is the point at which all of the antioxidant stabilizer has been consumed. These experiments are typically carried out isothermally at temperatures which exceed the melting point of the polymer sample (e.g. 180-210°C). As samples can be evaluated relatively quickly, this method is quite popular within the industry.

EXPERIMENTAL

Oven Exposure: All samples were exposed in a vented, forced draft oven equipped with a rotating carousel. Samples were suspended from a mylar-lined metal clip. On a given time interval the samples were evaluated for embrittlement by a 90° bend test. Failure was breaking of the sample upon bending around a model with a 5 mm radius. Alternately, failure of samples was the initiation of powder formation in the main body of the sample or around the bottom edges of the sample (i.e. not adjacent to the mounting device).

Oxidative Induction Time Determination: Oxidative induction times of 10 mil films were obtained at 200°C by Differential Scanning Calorimetry. Discs were cut from the film using a stainless steel paper punch (6.4 mm dia.) and placed in clean aluminum DSC pans. The samples heated in the DSC at 20°C/min to 200°C under a nitrogen atmosphere (100ml/min purge) and allowed to equilibrate at 200°C for approximately 30 seconds. The purge gas was then switched to oxygen and the samples were held at 200°C under O_2 until an exotherm was detected (approx. 10 mW full scale). The switch to an oxygen purge gas marks time zero, and the time to the extrapolated onset of the oxidation exotherm is reported. All contour maps were generated using a standard quadratic experimental design (residual SD = 4.03; replicate SD = 2.48).

RESULTS AND DISCUSSION

Correlation between OIT and Oven Aging: The results of a comparison between OIT and oven aging with carbon black containing HDPE are shown in Table 1 [9]. Clearly, the

Table 1 -- Comparison of test methods.

Stabilization[a]	OIT[b] @ 200°C (Minutes)	T50[c] @ 120°C (Days)
Base Alone	3.5	15
0.1% AO-1	45.0	109
0.1% AO-3	57.0	111
0.1% AO-4	48.0	226
0.1% HALS-1	6.0	217
0.1% HALS-2	4.0	199

[a]Base: High Density Polyethylene + 2.5 Carbon Black.
[b]OIT: Isothermal at 200°C, 100 ml/min O2, open Al pans.
[c]T50: Time to 50% retained tensile impact strength (1 mm plaque).

initial OIT values do not correlate to the time to 50% retained tensile impact strength after exposure at 120°C. Hindered phenolic antioxidants have been traditionally used both as OIT enhancers and thermal stabilizers (during processing and long term aging). Interestingly, the relative performance of these antioxidants changed significantly as the temperature dropped from 200 to 120°C. Hindered Amine Light Stabilizers (HALS) had very little impact on OIT, but performed superbly at 120°C. Figure 1 demonstrates the poor correlation between the two test methods.

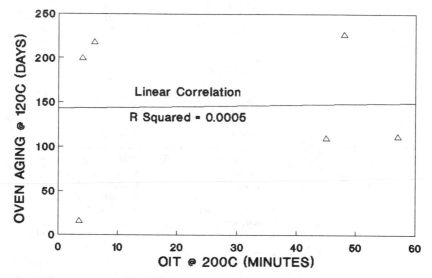

Fig. 1 -- OIT - oven agin correlation diagram - HDPE

A variety of structurally dissimilar hindered phenolic antioxidants in polypropylene were evaluated by OIT (200°C) and oven aging (150°C). The results of this study are summarized in Table 2. As was the case with HDPE, no

Table 2 -- Comparison of test methods.

Stabilization[a]	OIT[a] @ 200°C (Minutes)	Failure[c] @ 150°C (Days)
Base Alone	1.3	1
0.1% AO-1	6.7	58
0.1% AO-2	5.0	14
0.1% AO-4	8.4	46
0.1% AO-5	8.5	32
0.1% AO-6	8.5	22
0.1% AO-7	8.3	5
0.1% AO-9	1.7	2

[a]Base: PP homopolymer + 0.075% CaST + 0.1% P-1.
[b]OIT: Isothermal at 200°C, 100 ml/min O_2, open Al pans.
[c]Days to catastrophic Failure.

correlation appears to exist between oven aging and OIT
(Figure 2).

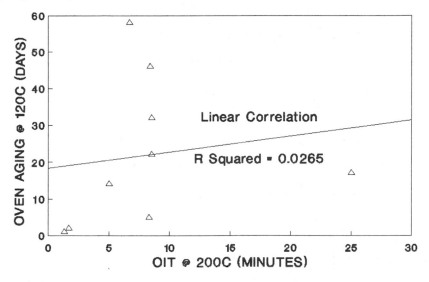

Fig. 2 -- OIT - oven aging correlation diagram - PP.

A study of HDPE insulation for telephone cables by
Bernstein and Lee [10] demonstrates the potential problems
associated with extrapolation of OIT data. Figure 3 shows
Arrhenius plots based on oven aging data and OIT data.
The response of oven aged samples in this plot is not
linear. As the test temperature (or application
temperature) decreases, the predominant stabilization
pathway may also change leading to a non-linear Arrhenius
plot. The stabilization mechanisms important in the
relatively high temperature ranges used in OIT may not be
relevant at lower temperature.

The potential consequences of these differences are
described in Figure 3. Using the OIT method, a given
formulation has a projected life of 189 years at 70°C,
whereas oven aging data indicates an expected life time of
only 12 years! This difference is further exaggerated at
lower temperatures, i.e. from a predicted durability of
greater than 55,000 years (based on OIT) to down to 304
years (based on oven aging).

<u>Effect of Antioxidant Concentration</u> OIT values were
measured at 190°C for a series of melt extruded films
containing various levels of antioxidants AO-1 and AO-2.
The results are plotted in Figure 4. Although the
relationship between OIT and antioxidant concentration is
not strictly linear throughout the concentration range,
OIT would appear to function well for quality control
purposes. Plotting days to brittleness (135°C) as a
function of antioxidant concentration yields, a fairly

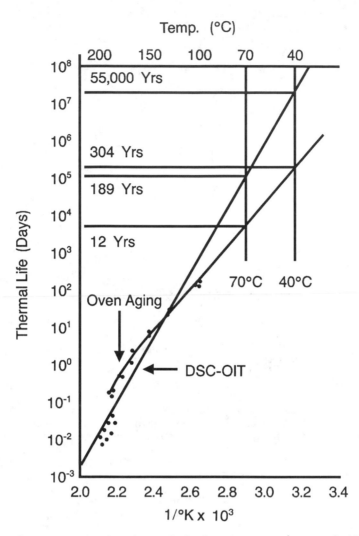

Fig. 3 -- Arrhenius plot for oven aging and OIT for
HDPE insulated cables[10].

linear relationship through most of the concentration
range as shown in Figure 5.

Effect of Other Additives: Other classes of stabilizers
can have a significant influence on OIT values. Figure 6
contains a contour map showing the effect on OIT (minutes
to onset of exotherm) of varying the concentrations of
both AO-1 and a phosphite, P-1. Both AO-1 and P-1 are
effective as OIT improvers. The effect appears to be
additive, where the maximum OIT is achieved by maximizing
the concentrations of both compounds. This interaction is
interesting in light of the fact that phosphites are

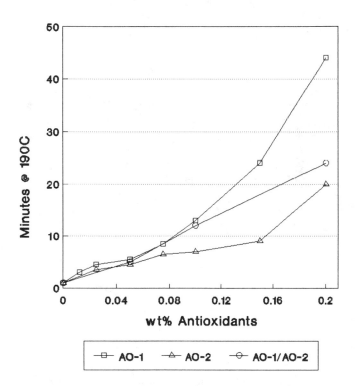

Fig. 4 -- Effect of antioxidant concentration on OIT.
Base: PP homopolymer + 0.1% CaSt.

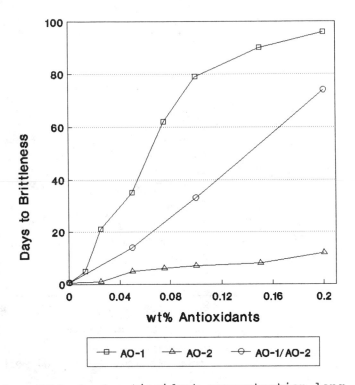

Fig. 5 -- Effect of antioxidant concentration long
          term oven aging.
          Base: PP homopolymer + 0.1% CaSt;
          film (120 um).

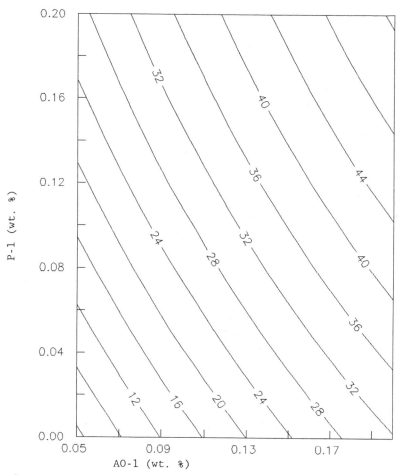

Fig. 6 -- Effect of P-1 and AO-1 concentration on OIT
(minutes).   Base contains HDPE + 2.5%
carbon black.

generally not considered to significantly improve long
term thermal stability.

Figure 7 shows the effect of thioether/phenolic
antioxidant combinations. As in the case of phosphites,
thioethers can significantly boost OIT values particularly
when used in combination with phenolic antioxidants such
as AO-1.

The effect on OIT of HALS-1 in combination with AO-1 is
shown in Figure 8. As previously demonstrated, OIT
increases with increasing AO-1 concentration. HALS-1 has
little impact on the OIT value. However, this is not the
case in formulations containing a thiosynergist such as
DSTDP (distearyl-thio-dipropionate). At very low

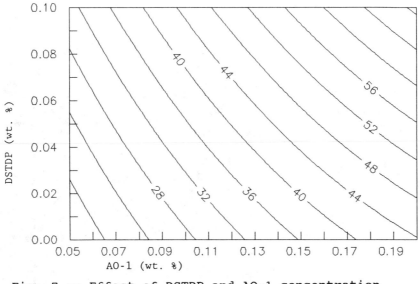

Fig. 7 -- Effect of DSTDP and AO-1 concentration
          on OIT (minutes). Base contains
          HDPE + 0.10% P-1.

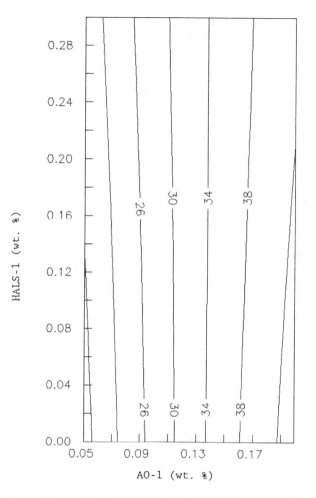

Fig. 8 -- Effect of HALS-1 and AO-1 concentration
on OIT (minutes). Base contains
HDPE + 2.5% carbon black + 0.1% P-1.

concentrations of HALS-1, OIT increases with DSTDP
concentration as shown in Figure 9.   Beyond a HALS-1

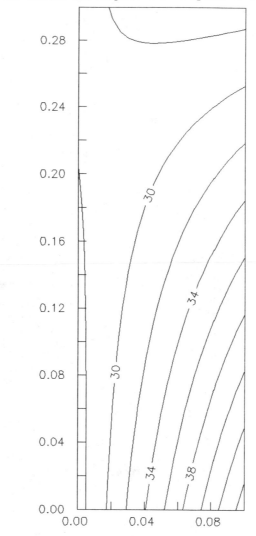

DSTDP (wt. %)

Fig. 9 -- Effect of HALS-1 and DSTDP concentration
    on OIT (minutes).   Base contains HDPE +
    2.5% carbon black + 0.1% AO-1 + 0.1% P-1.

concentration of 0.2%, DSTDP fails to have any positive
effect on OIT.   In these types of formulations, OIT will
have little use even as a quality control method.

In polypropylene, oven aging data shows a similar
interaction, as indicated by the data in Table 3.

Table 3 -- 125°C LTHA of carbon black filled PP plaques.

Light Stabilizer[a]	Hours to Failure[b] @ 125°C	
	0.1% AO-1 + 0.1% P-1	0.1% AO-1 0.3% DSTDP
Base Alone	600	1,900
0.10% HALS-2	1,200	1,100
0.50% HALS-2	5,000	4,700
0.10% HALS-1	2,000	2,000
0.10% HALS-1	5,000	4,000

[a]Base: PP homopolymer + 0.1% CaST + 2.5% carbon black.
[b]Hours to catastrophic failure.

Addition of a HALS to a polypropylene sample can increase its heat stability over seven-fold. As observed in the OIT method with HDPE samples, addition of a thioether can actually reduce the heat stability of these systems. The extent of this interaction is somewhat dependent upon the type of HALS used. In these carbon black filled polypropylene samples, HALS-1 appears to have less interaction with the thioether than does HALS-2.

The various mechanisms [11] by which differing stabilizers function can have a profound effect on their performance in a particular test. HALS appear to rapidly increase their heat stabilizing performance as temperatures decrease. Therefore, HALS appear most active at temperatures below 150°C. Conversely, thioether and phosphites provide their best performance at higher temperatures. Liner specifications requiring testing at high temperatures may result in products that are designed to pass a particular test but are not optimized for the actual lower temperature application.

One additional consideration in choosing a testing temperature is the polymer melting point. In raising the temperature beyond the polymer melting point, the reaction kinetics of key processes may significantly change. As the polymer is no longer a solid, reaction will be occurring in solution rather than through solid/solid or solid/gas interfaces.

Polymer morphology is known to play an important role in degradation. Amorphus regions in the polyolefin are believed to undergo degradation much more readily than crystalline regions. Destruction of the molecular structure of the solid through melting may significantly impact the results of the study.

CONCLUSIONS

OIT is a convenient tool for determining antioxidant level. In many cases it can be reliably used for quality control type evaluations. It should be noted, however, that in certain cases (e.g. thioether/HALS formulations) use of this method is not appropriate.

It is obvious from the data presented in this paper that OIT does not correlate to testing carried out at lower temperatures (oven aging). As testing temperatures decrease and approach in-use conditions, accuracy in predicting liner durability will certainly improve. However at some point testing times will become prohibitively long and therefore the test temperature chosen must necessarily be a compromise between rapid testing times and realistic results.

As new accelerated test methods are developed, careful correlation studies between the proposed method and low temperature oven aging must be completed before the method can be adopted. This is particularly important for methods which involve melting the polymer sample.

ACKNOWLEDGEMENTS

The contributions and research efforts of the Polyolefin Additives Laboratories in both Ardsley, New York and Basel, Switzerland are gratefully acknowledged. In particular, I would like to thank Bernice Marriot (Ardsley) and François Gugumus (Basel) for their valuable contributions. Special appreciation is also extended to the CIBA-GEIGY Corporation for permission to use the data presented.

REFERENCES

(1)   White, D.F.; Verschoor, K.L. <u>Geotechnical Fabricsreport</u> **1989**, 7(3), 16.

(2)   Ford, J. E., <u>Fourth International Conference on Polypropylene Fibres and Textiles</u>, Nottingham, UK, September **1987**.

(3)   Tisinger, L.G. <u>Geotechnical Fabricsreport</u> **1989**, 7(3), 22.

(4)   Ramaswamy, S. D.; Rathor, M. N., <u>Fourth International Conference on Polypropylene Fibres and Textiles</u>, Nottingham, UK, September **1987**.

(5)   ASTM Committee D-35 "ASTM Standards on Geotextiles", Philadelphia, PA, **1988**.

(6)   Forsman, J.P., _S.P.E. Tech. Papers_, **1964**, _10_, VIII-2.

(7)   Still, R.H., _Developments in Polymer Degradation-1_, Applied Science Publishers, London, **1977**.

(8)   Billingham, N.C.; Bott, D.C.; Manke, A.S., _Developments in Polymer Degradation-3_, Applied Science Publishers, London, **1981**.

(9)   Gugumus,F., _Polymer Degradation and Stability_, **1989**, 24, 289.

(10) Bernstein, B.S.; Lee, P.N., _Proc. 24th Int. Wire and Cable Symp._, November, **1975**.

(11) Gray, R.L.; _ACS Symposium Series_, in press.

APPENDIX I

ABBREVIATIONS:

AO-1

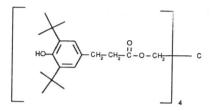

(Irganox 1010)

AO-2

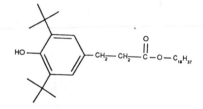

(Irganox 1076)

AO-3

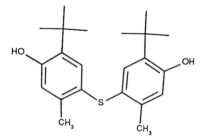

(Santanox R)

AO-4

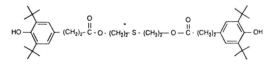

(Irganox 1035)

APPENDIX I
(con't)

ABBREVIATIONS:

AO-5

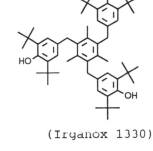

(Irganox 1330)

AO-6

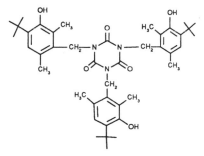

(Cyanox 1790)

AO-7

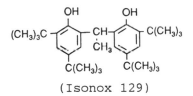

(Isonox 129)

AO-8

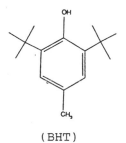

(BHT)

APPENDIX I
(con't)

ABBREVIATIONS:

DSTDP

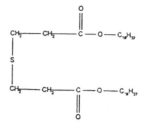

(Irganox PS-802)

P-1

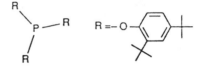

(Irgafos 168)

HALS-1

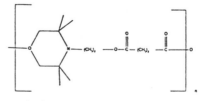

(Tinuvin 622)

HALS-2

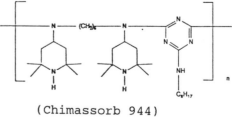

(Chimassorb 944)

Henry E. Haxo, Jr.

DETERMINING THE TRANSPORT THROUGH GEOMEMBRANES OF
VARIOUS PERMEANTS IN DIFFERENT APPLICATIONS

REFERENCE: Haxo, H. E., "Determining the Transport Through
Geomembranes of Various Permeants in Different Applications,"
Geosynthetic Testing for Waste Containment Applications,
ASTM STP 1081, Robert M. Koerner, editor, American Society
for Testing and Materials, Philadelphia, 1990.

ABSTRACT: The widely different uses of geomembranes as barriers to the transport and migration of different gases, vapors,
and liquids under different service conditions require a determination of permeability by test methods appropriate to the
service. Geomembranes are nonporous and homogeneous materials
that permeate gases, vapors, and liquids on a molecular scale
by dissolution in the geomembrane and diffusion through the
geomembrane. The rate of transmission of a given species,
whether as a single permeant or in mixtures, is driven by its
chemical potential or concentration gradient. Various methods
to assess the permeability of geomembranes to single component
permeants, such as individual gases, vapors, and liquids are
described and data are presented. In addition, various test
methods for the measurement of permeation and transmission
through geomembranes of individual species in complex mixtures
such as waste liquids are described and data are presented.

KEYWORDS: Permeability, gas transmission, water vapor transmission, transport of chemical species, polymeric geomembranes, organic vapor transmission, diffusion of gases and vapors, transport of ions, reservoirs, waste disposal, leachate, barriers,
flexible membrane liners (FMLs).

INTRODUCTION

The principal characteristic of geomembranes is their intrinsi-
cally low permeability to a broad range of gases, vapors, and liquids,
both as single-component fluids and as complex mixtures of many constituents. As low permeable materials, geomembranes are being used
in a wide range of engineering applications in geotechnical, environmental, and transportation areas as barriers to control the migration
of mobile fluids and their constituents. The range of potential
permeants is broad and the service conditions can differ greatly.

The transmission of various species through a geomembrane is
subject to many factors which must be assessed in order to be able

Dr. Henry E. Haxo is President of Matrecon, Inc., 815 Atlantic
Avenue, Alameda, CA 94501.

to predict its effectiveness for a specific service. Permeability measurements are affected by test conditions and measurements made by one method cannot be translated from one application to another. A wide variety of permeability tests have been devised to measure the permeability of polymeric materials; however, only a limited number of these procedures have been applied to geomembranes. Test procedures should be selected to reflect actual service requirements as closely as possible.

This paper discusses the mechanism of permeation of mobile chemical species through geomembranes and the permeability tests that are relevant to various types of applications and permeating species. Specific tests for the permeability of geomembranes to both single-component permeants and multi-component fluids that contain a variety of permeants are described and discussed, and data on the permeability of the various geomembranes obtained in the various procedures are presented.

MECHANISM OF PERMEATION OF GEOMEMBRANES

Even though geomembranes are nonporous and cannot be permeated by liquids as such, gases and vapors of liquids can permeate a geomembrane on a molecular level. Thus, even if a geomembrane is free of holes, some components of the contained fluid can permeate and might escape the containment unit.

The basic mechanism of permeation through geomembranes is essentially the same for all permeating species. The mechanism differs from that through porous media, such as soils and concrete, which contain voids that are connected in such a way that a fluid introduced on one side will flow from void to void and emerge on the other side; thus, a liquid can flow advectively through the voids and carry dissolved species.

Overall rate of flow through saturated porous media follows Darcy's law which states that the flow rate is proportional to the hydraulic gradient, as is shown in the following equation:

$$Q = kiA \tag{1}$$

where

Q = the rate of flow,

k = a constant (Darcy's coefficient of permeability),

A = the total inside cross-sectional area of the sample container, and

i = the hydraulic gradient.

With most liquids in saturated media, the flow follows Darcy's law; however, the flow can deviate due to interactions between the waste liquid and the surface of the soil particles. These interactions become important in the escape of dissolved species through a liner system in a waste facility. Dissolved chemical species [1], either organic or inorganic, not only can permeate a low permeability porous medium advectively (i.e. the liquid acts as the carrier of the chemical species), but also by diffusion in accordance with Fick's two laws of diffusion.

Even though polymeric geomembranes are manufactured as solid homogeneous nonporous materials, they contain interstitial spaces between the polymer molecules through which small molecules can diffuse. Thus, all polymeric geomembranes are permeable to a degree. A permeant migrates through the geomembrane on a molecular basis by an activated diffusion process and not as a liquid. This transport process of chemical species involves three steps:

1. The solution or absorption of the permeant at the upstream surface of the geomembrane.
2. Diffusion of the dissolved species through the geomembrane.
3. Evaporation or desorption of the permeant at the downstream surface of the geomembrane.

The driving force for this type of activated permeation process is the "activity" or chemical potential of the permeant which is analogous to mechanical potential and electrical potential in other systems. The chemical potential of the permeant decreases continuously in the direction of the permeation. Concentration is often used as a measure of the chemical potential.

In the transmission of a permeant through a geomembrane, Step 1 depends upon the solubility of the permeating species in the geomembrane and the relative chemical potential of the permeant on both sides of the interface. In Step 2, the diffusion through the geomembrane involves a variety of factors including size and shape of the molecules of the permeating species, and the molecular characteristics and structure of the polymeric geomembrane. A steady state of the flow of the constituents will be established when, at every point within the geomembrane, flow can be defined by Fick's first law of diffusion:

$$Q_i = - D_i \frac{dc_i}{dx} \tag{2}$$

where

Q_i = the mass flow of constituent "i" in g cm^{-2} sec^{-1},

D_i = the diffusivity of constituent "i" in cm^2 sec^{-1},

c_i = the concentration of constituent "i" in g cm^{-3}, and

x = the thickness of the geomembrane in cm.

It should be noted that the concentration of constituent "i" referred to in Fick's law is within the mass of the geomembrane.

Step 3 is similar to the first step and depends on the relative chemical potential of the permeant on both sides of the interface at the downstream geomembrane surface.

TABLE 1 -- Permeability of Polymeric Geomembranes to Selected Gases at 23°C, Determined in Accordance with ASTM D1434, Procedure V [3]

| Geomembrane description | | | | Gas transmission rate mL(STPb)/m²·d·atm | | | Gas permeability coefficient ($\overline{P}$), barrerc | | |
Base Polymer	Density	Thicknessa, mm	Compound typed	CO2	CH4	N2	CO2	CH4	N2
Butyl rubber	...	1.60	XL	512	120	19.7	12.5	2.92	0.480
Chlorosulfonated polyethylene	...	0.82R	TP	122	21.6	26.2	1.52	0.27	0.33
	...	0.86	TP	418	124	27.1	5.47	1.62	0.36
Elasticized polyolefin	...	0.58	CX	1450	280	125	12.8	2.47	1.10
Ethylene propylene rubber	...	0.89R	TP	2720e	...	...	36.8e	...	...
	...	0.90	XL	5260	1400	314	72.0	19.2	4.30
Neoprene	...	0.90	XL	716	80.9	31.1	9.81	1.11	0.43
Polybutylene	...	0.71	CX	818	248	62.3	8.84	2.68	0.67
Low-density polyethylenef	0.921	0.25	CX	6180e	1340e	...	23.5e	5.10e	...
Linear low-density polyethylene	0.923	0.46	CX	1370	322	...	9.59	2.25	...
High-density polyethylene	0.945	0.61	CX	729	138	...	6.77	1.28	...
	0.945	0.86	CX	467	104	...	6.11	1.36	...
Polyvinyl chloride	...	0.25	TP	7730e	1150e	...	29.4e	4.38e	...
(plasticized)	...	0.49	TP	3010	446	108	22.4	3.32	0.81
	...	0.81	TP	2840e	285e	...	35.0e	3.51	...
Polyester elastomerg	...	0.022	CX	357	...	...	0.119	...	...

aR = fabric-reinforced.

bSTP = standard temperature and pressure.

cOne barrer = 10^-10 mL(STP)·cm/cm²·s·cm Hg.

dXL = crosslinked; TP = thermoplastic; CX = semicrystalline.

eMeasured at 30°C.

fNatural resin (no carbon black).

gThis sample is NBS Standard material 1470. The determination was made at 15.0 psi, under which condition the NBS Certified CO2 transmission rate was calculated to be 338 mL(STP)/m²·d·atm.

Chemical potential is an idealized concept which indicates the direction in which the permeation will go, i.e. from high to low potential. To use concentration directly to replace chemical potential requires the individual molecules of the permeating species to neither interact with each other nor with the membrane they are permeating. This condition approximately exists when a permanent or a noncondensable gas, such as oxygen, nitrogen, or helium, permeates a membrane. However, the individual molecules of organic species can interact with each other and with the polymer to increase solubility of the species in the geomembrane, and as a result partition to the geomembrane.

PERMEABILITY OF GEOMEMBRANES TO SINGLE-COMPONENT FLUIDS

Many of the applications of geomembranes are for barriers to the permeation of single-component permeants, i.e. a single gas, vapor, or liquid. With respect to water, such applications include reservoir liners, moisture vapor transmission barriers, floating covers for reservoirs, canal liners, tunnel liners, methane barriers, and liners for secondary containment. Various tests that are appropriate for assessing barriers to the permeation of different types of single-component fluids are discussed.

Permeability of Geomembranes to Single Gases

For such applications as linings for waste disposal facilities and methane barriers, the permeability to gases is important in geomembrane selection. The permeability of geomembranes can be assessed by measurement of the volume of the gas passing through the geomembrane under specific conditions or by measurement of the increase in pressure on the evacuated downstream side. Both methods are described in ASTM D1434 [3].

The volumetric method was used by Haxo et al [4], to measure the permeability of a wide range of geomembranes to methane, carbon dioxide, and nitrogen. In this procedure, the geomembrane is in contact with the gas on both sides, i.e. on the upstream side at a pressure greater than atmospheric and on the downstream side at atmospheric pressure to yield a concentration gradient and diffusion of the gas in the geomembrane.

Some results obtained by Haxo et al [5] are presented in Table 1 for measurements on 15 different commercial geomembranes and one reference film. Data are presented on the gas transmission rates for specific geomembranes and the normalized data in barrers for the permeability coefficients. The data show the great range in permeability coefficients among the geomembranes, including compounding variations, and among the three gases.

Other variables that should be considered in assessing the gas transmission rate (GTR) of a given gas include thickness and such test conditions as temperature and pressure. The GTR and the permeability coefficient as affected by thickness are shown in Table 2 for methane at 23°C through HDPE of various thicknesses. The effect of temperature on permeability is illustrated in Figure 1; the values decrease linearly with the reciprocal of absolute temperature.

TABLE 2 -- Effect of Thickness on the Permeability
Coefficient of HDPE to Methane[a]

| | Thickness, mm | | | |
Permeability property	0.64	0.85	1.27	2.54
Gas transmission rate, $g \cdot m^{-2} \cdot d^{-1}$	150	87	47	20
Permeability coefficient, barrers[b]	1.47	1.13	0.94	0.778

[a]Based on Figure 4-21 [5].

[b]One barrer = 10^{-10} mL(STP)$\cdot$cm/cm$^2 \cdot$s$\cdot$cm Hg.

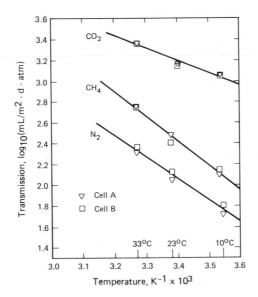

FIG. 1 -- Permeability of elasticized polyolefin to CO_2,
CH_4, and N_2 as a function of temperature.

Permeability of Geomembranes to Water

Permeability to Moisture Vapor: For applications such as reservoir covers and moisture barriers, permeability to moisture vapor can be measured by a variety of methods that reflect the service conditions. Determinations can be made by measuring the change in weight of a small cup that contains either a small amount of distilled water or a desiccant and is sealed at the mouth with a specimen of the geomembrane, e.g. ASTM E96 [3].

Test results obtained by Haxo et al [5] in accordance with this method are presented in Table 3. The number of geomembranes of each type tested and the ranges of thicknesses as well as the average values and ranges of test values are shown.

In these tests, ASTM E96, Water Method, Procedure BW [3], was used; aluminum cups were partially filled with deionized (DI) water and covered by the respective geomembrane. The cups were kept in an inverted position so that water was in contact with the samples and were placed in an environment where the relative humidity (RH) was 50%; a standard flow of air over the geomembrane specimen was maintained.

Permeability to Water: Under a head of water comparable to that encountered in a water reservoir, the pressure on the surface of a geomembrane can cause a small transmission of water through the geomembrane. Various measurements of water permeating geomembranes have been made in which pressure has been applied across a geomembrane with the water on the downstream side at atmospheric pressure [6, 7]. The amount of aerated water that was transmitted through the membrane was measured on the downstream side. This type of permeability test applies only to water or waters of equal concentration on both sides of the geomembrane. A brine or a waste liquid on the upstream side could reverse the direction of permeation of water due to osmotic pressure. The water permeabilities of several geomembranes determined in this type of test are reported as hydraulic conductivity coefficients at several upstream pressures (see Table 4).

Permeability of Geomembranes to Organics

Permeability to Organic Vapors: The moisture vapor transmission test can also be used to assess the permeability of various membranes to solvent vapors. In this case, the cup that is used in the moisture vapor transmission test is exposed with the solvent vapor contacting the membrane. The vapor concentration inside the cup is that of the vapor pressure at the test temperature and the concentration outside the cup is essentially zero. Therefore, the vapor pressure gradient is the vapor pressure of the solvent at the temperature of test.

Modifying the test procedure and calculating the transmission rates, Haxo et al [4, 5] obtained the results presented in Table 5 for five solvents with five semicrystalline geomembranes and polytetrafluoroethylene. The data show significant differences among the geomembranes and among the solvents.

These limited results show particularly low permeability of the geomembranes to the polar solvents, methyl alcohol and acetone, in comparison to the hydrocarbon and chlorinated solvents. The effects of thickness on vapor transmission rate and the permeability coefficients are shown for HDPE. The transmission rate for the thicker specimen (2.60 mm) was approximately 30% of that of the thinner specimen (0.80 mm); however, the vapor permeability coefficients which normalize the data are essentially equal. All of the transmission data presented are based on saturated vapors and would be

TABLE 3 -- Moisture Vapor Permeability of Geomembranes and Films Based on Different Polymers Measured in Accordance with ASTM E96a, Procedure BW [3]

Base polymer	Type of compound[b]	Number of membranes in average	Range of thickness, mm	Moisture vapor permeability (10^{-2} metric perm·cm)	
				Average	Range of values
Butyl rubber	XL	3	0.85-1.18	0.081	0.016-0.170
Chlorinated polyethylene	TP	11	0.53-0.97	0.485	0.213-1.27
Chlorosulfonated polyethylene	TP	10	0.74-1.07	0.460	0.234-0.845
Elasticized polyolefin	CX	2	0.61-0.72	0.090	0.083-0.097
Epichlorohydrin rubber	XL	2	1.16-1.65	22.3	22.2-22.4
Ethylene propylene rubber	XL	6	0.51-1.70	0.260	0.131-0.367
Neoprene	XL	6	0.51-1.59	0.354	0.147-0.517
Nitrile rubber	TP	1	0.76	3.98	...
Polybutylene	CX	2	0.19-0.69	0.064	0.055-0.072
Polyester elastomer	CX	2	0.20-0.25	1.99-10.6	1.99-10.6
Polyethylene (low density)	CX	1	0.76	0.041	...
Polyethylene (high density)	CX	2	0.80-2.44	0.014	0.013-0.014
Polyethylene (high density) - alloy	CX	1	0.86	0.039	...
Poly (ethylene vinyl acetate)	TP	1	0.53	0.76	...
Polyvinyl chloride	TP	8	0.28-0.79	1.30	0.77-1.45
Polyvinyl chloride (elasticized)	TP	2	0.91-0.97	2.10	1.79-2.40
Polyvinyl chloride (oil-resistant)	TP	3	0.79-0.84	3.08	2.60-3.35
Polyvinylidene chloride (Saran)	TP	1	0.013	0.0070	...
Polytetrafluoroethylene (Teflon)	CX	1	0.10	0.0021	...

aASTM E96-80, Procedure BW: Inverted water method at 23°C; 50% humidity on downstream side. Permeance in metric perms = g m^{-2} d^{-1} (mm Hg)$^{-1}$ = WVT/Δp in mm Hg, where Δp = the vapor pressure difference = 10.53 mm Hg (at 23°C and 50% humidity on downstream side). Permeability in metric perms·cm = permeance x thickness of geomembrane in cm.

bXL = crosslinked; TP = thermoplastic; CX = semicrystalline.

considerably lower in landfill environments where the vapor concentrations are less than saturated.

TABLE 4 -- Coefficient of Permeability to Deaerated
Water (Hydraulic Conductivity) of Several Geomembranes
at Various Pressures[a]

| | Hydraulic pressure, kPa | | | |
Base polymer	100	400	700	1000
Butyl rubber	22	8.0	4.1	2.8
Chlorosulfonated polyethylene	300	120	77	60
Ethylene propylene rubber	115	57	34	25
Polyvinyl chloride	86	37	20	10

[a]Data are reported in 10^{-15} m·s^{-1}; based on averaged data taken from Giroud [6].

TABLE 5 -- Permeability of Polymeric Geomembranes to
Various Solvents, Measured in Accordance with ASTM E96,
Procedure B (Modified to Test Solvents) [3]

Polymer[a]	ELPO	HDPE	HDPE	HDPE-A	LDPE	PB	Teflon
Average thickness, mm	0.57	0.80	2.62	0.87	0.75	0.69	0.10
Type of compound	CX	CX	CX	CX	CX	CX	CX
SVT, g m^{-2} d^{-1}							
Methyl alcohol	2.10	0.16	...	0.50	0.74	0.35	0.34
Acetone	8.62	0.56	...	2.19	2.83	1.23	1.27
Cyclohexane	7.60	11.7	...	151	161	616	0.026
Xylene	359	21.6	6.86	212	116	178	0.16
Chloroform	3230	54.8	15.8	506	570	2120	20.6
Solvent vapor permeability[b], 10^{-2} metric perms·cm							
Methyl alcohol	0.11	0.01	...	0.04	0.05	0.02	0.003
Acetone	0.23	0.02	...	0.09	0.10	0.04	0.006
Cyclohexane	0.49	1.05	...	14.7	13.6	47.8	0.0003
Xylene[c]	292	24.6	25.6	262	124	175	0.002
Chloroform	103	2.46	2.32	24.6	24.0	82.2	0.12

[a]ELPO = elasticized polyolefin; HDPE = high-density polyethylene;
HDPE-A = high density polyethylene alloy; LDPE = low-density
polyethylene; PB = polybutylene.

[b]The median thickness value was used to calculate the permeability.

Another test method that can be used for measuring permeability to organic vapors is ASTM F739 [3], which is used to measure the resistance of protective clothing materials to the permeation of liquids or gases. In this method, an analytical detection system is used to measure the time to breakthrough of the permeant and the equilibrium rate of permeation.

Permeability to Organic Liquids: For those applications in which geomembranes will be contacted by organic liquids, such as liners for tanks and secondary containment, it is necessary first to determine the compatibility of the specific membrane with the specific organic that is to be contained. This is necessary because of the potential swelling of the geomembrane which can change the permeability.

Compatibility testing has been used in the rubber and plastics industries for assessing compatibility of coatings and lining materials for equipment and pipes. A test commonly used for this purpose is ASTM D471. Once compatibility has been demonstrated, tests such as ASTM E96 or ASTM D841, in which the solvent contacts the specimen, can be used and treated in a similar fashion to modified ASTM E96 [3].

PERMEABILITY OF GEOMEMBRANES TO MULTI-COMPONENT FLUIDS

Many of the applications of geomembranes as barriers involve contact with multi-component fluids, e.g. mixtures of gases, liquids, and aqueous solutions of salts and/or organics. The most complex of such mixtures are probably leachates from waste disposal facilities. In considering geomembranes for these applications, one must recognize the great difference in the rates of different chemical species and recognize that the rates depend on solubility, diffusibility, and concentration gradient across the membrane; also, the permeating species may interact differently with each other and with the geomembrane. Though some of the basic test methods described for single-component permeants can be used, they must be supplemented in most cases by a means of identifying species that have permeated the membrane. The analysis of the permeant on the downstream side is needed because of the selective nature of polymeric membranes which results in different transmission rates for different chemical species. Such analytical tools as gas chromatography (GC) or GC mass spectrography (GCMS) for organics and atomic absorption and analyses the inorganics are often used to detect, identify, and quantify the permeant.

Permeability of Geomembranes to Mixtures of Gases

In many of the applications as barriers to the migration of gases, the geomembrane will encounter a mixture of two or more gases, which, due to the permselectivity of the geomembrane, will permeate at different rates. Data on the permeability to mixtures of CO_2 and N_2 of films of several of the polymers that are used in geomembranes are presented in Table 6. The permeabilities of the films to the two gases vary over a range of about 5,000, yet the ratios of their permeabilities in the respective films remain relatively constant, i.e. about 2. GC or GCMS must be used to analyze the permeating mixtures.

TABLE 6 -- Permeability of Different Polymeric Films
to a Mixture of Carbon Dioxide and Nitrogen at 30°C [2]

Film	P_{CO_2}a x 10^{11}	P_{N_2}a x 10^{11}	Ratio
Saran	0.29	0.0094	30.9
Polyester[b]	1.53	0.05	30.6
Butyl rubber	518	3.12	17.4
Neoprene	250	11.8	21.1
Polyethylene	352	19	18.5
Natural rubber	1310	80.8	16.3

acm^3(STP)·cm/cm^2·sec·cm Hg.

bMylar, Du Pont.

Permeability of Geomembranes to Aqueous Solutions of Inorganic Salts

Geomembranes are being used to line wastewater and solid waste
storage and disposal facilities which contain aqueous solutions of inorganic salts, e.g. leachates from coal-fired power plant wastes. In
this example, a geomembrane functions as a semi-permeable barrier to
the migration of inorganic salts. The permeability of the geomembrane
to ions can be measured by separating the solution containing the ions
from DI water and measuring, as a function of time, the electrical
conductivity (EC) of the DI water, or by measuring the concentration
of the specific ions. If the geomembranes can be fabricated into
pouches, a pouch-type test can be used to assess the permeability of
the ions and the water in the liquid [5].

As an example of the measurement of the permeation of ions and
water, pouches of PVC were filled with 5 and 10% solutions of lithium
chloride and placed in DI water. The EC of the outer water exhibited
almost no change during exposures of up to 600 days. However, as the
result of osmotic pressure, the pouches gained in weight (see Figure
2). These results show that the ions did not permeate the pouch
walls but the water permeated into the pouch from the outer DI water.
Because lithium ions, which are not commonly found in impoundment
environments, do not permeate a geomembrane but would pass through a
hole, they are potentially useful as a tracer for leaks in a liner.

Permeability of Geomembranes to Mixtures of Organics

For applications of geomembranes that contact mixtures of
organics which might affect the geomembrane, such as in secondary
containment and tanks, compatibility and permeability tests of the
geomembranes with the potential mixture should be performed. Testing
of a geomembrane with an individual component of a mixture cannot
reflect the potential interaction of the organics and their combined
effects on the geomembrane.

August and Tatsky [8] measured the transmission rates of each of
6 solvents from an equivolume fraction mixture through a 1.0 mm HDPE
geomembrane. They used an apparatus consisting of two compartments
separated by the geomembrane. The upper compartment contained the

solvent mixture, the composition of which was held constant, and the
lower compartment was partially evacuated. The permeating vapors were
analyzed by gas chromatography. The data presented in Table 7 show
the great difference in the rates among the solvents, though the order
is similar to that obtained on single solvents.

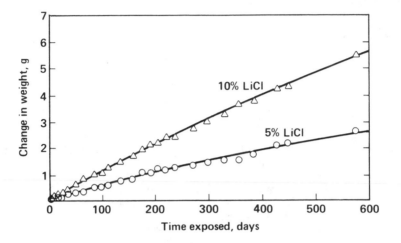

FIG. 2 -- Weight change of PVC pouches containing 5 and
          10% aqueous solutions of LiCl during immersion
          in DI water.

TABLE 7 -- Permeation Rates of the
Components of a Mixture of Organics Through
a 1.0 mm HDPE Geomembrane [8]

Organic	Permeation rate, $g\ m^{-2}\ d^{-1}$
Trichloroethylene	9.4
Tetrachoroethylene	8.1
Xylene	3.0
Isooctane	0.8
Acetone	1.4
Methanol	0.7
Total	23.4

A similar experiment was performed by Haxo et al [4] in which the
transmission rates of the components of a mixture containing equal
volumes of 6 solvents through a 0.5 mm elasticized polyolefin geomembrane were monitored. This testing was performed in accordance with a
procedure based on ASTM E96, Procedure B [3]. The all-metal cups were
stored in a headspace can in an upright position so that the vapors

contacted the geomembrane specimens. The amounts of the solvent vapors that permeated the geomembranes were monitored in the headspace by gas chromatography. Table 8 presents the data obtained.

Results of these two studies indicate that strong selectivity by the geomembrane causes very different permeation rates for components of mixtures.

Permeability of Geomembranes to Aqueous Solutions of Organics

As a barrier material for waste storage and disposal facilities, geomembranes will probably contact dilute aqueous solutions of organics, e.g. leachates and waste liquids. Due to the differences in the solubility of individual organics in different geomembranes and in the partitioning coefficients between water solutions and the geomembranes, a considerable difference in the permeation rate of a given organic through a geomembrane compared with that obtained on the individual organic can be observed.

August and Tatzky [8] studied the permeation of organics in dilute aqueous solutions through a variety of geomembranes. The permeation rates of various pure organics and dilute solutions (0.1 to 0.001 weight percent) of the same organics through a 1.0 mm HDPE geomembrane were compared. August and Tatsky showed that the permeation of organics from a dilute solution can be substantially higher than would be expected from the reduced concentration. For example, even though the ratio between the concentrated toluene and the dilute solution was 1000:1, the ratio between permeation rates through the HDPE geomembrane was 20:1. These results indicate that significant quantities of an organic can permeate through a geomembrane due to selective permeation, even when the organics are present at a low concentration.

TABLE 8 -- Transmission of Solvent Mixtures Through a 0.5 mm Elasticized Polyolefin Geomembrane at 23°C

Time, h	Weight, % of original solvent remaining					
	Methanol[a]	MEK[b]	n-Heptane	TCA[c]	Toluene	Total[d]
0	17.6	18.0	15.2	29.8	19.3	100.0
22	≥17.3	16.6	12.6	27.0	17.2	90.8
70	≥16.8	13.8	7.9	21.8	13.2	73.7

[a]The methanol loss was below the analytical detection limit of the GC column. These data are based on a limiting value, the lower detection limit.

[b]MEK = methyl ethyl ketone.

[c]TCA = trichloroethane.

[d]The component values do not add exactly to the "total" value; see footnote "a". Additional errors were generated by manually integrating the loss rate data. The maximum error is 2.3%.

In an experiment performed by Haxo and Lahey [9], a closed apparatus consisting of three compartments separated by geomembranes was used to assess the permeation of organics from dilute aqueous solutions through polyethylene geomembranes. The middle compartment was partially filled with the solution, and DI water was placed in the bottom compartment. Thus, the organics could either volatilize into the airspace above the solution and then, permeating through the top geomembrane, enter the top compartment, or they could permeate the lower geomembrane into the bottom compartment. Septums were incorporated in each of the three compartments for withdrawing samples for GC analysis from the aqueous and airspace zones. After the apparatus was dismantled, the two geomembranes were analyzed by headspace GC.

The three-compartment apparatus simulated the configuration of a covered landfill, i.e.:

1. The airspace in the top compartment simulated the airspace over a "cover" liner. The geomembrane specimen between the top and middle compartments simulated a "cover" liner.
2. The airspace in the middle compartment simulates the headspace above a waste liquid, and the dilute solution containing organics serves as the waste liquid. The geomembrane specimen between the middle and bottom compartments simulates the service conditions of a bottom liner.
3. The airspace and the DI water in the bottom compartment simulate, respectively, pore spaces in the soil and the groundwater.

In an experiment to assess the distribution of organics among water, air, and a geomembrane and to assess the permeation of organics through a geomembrane, a dilute aqueous solution of toluene and trichloroethylene (TCE) was placed in the middle compartment of the test apparatus. An 0.84 mm linear low-density polyethylene (LLDPE) geomembrane separated the three compartments.

The middle compartment was filled with 500 mL of the dilute aqueous solution of toluene and TCE in DI water. The zones containing water or vapor were sampled and analyzed periodically by GC to track the changes in concentrations in the airspaces and water zones. After 256 hours, when the concentrations in these zones appeared to approach constant values and equilibrium had been reached, the apparatus was dismantled and the geomembranes were removed and analyzed by headspace GC to determine the concentrations of the organics in the membrane layers. Table 9 reports the results of the analyses of all the zones at the end of the exposure (256 hours) as averages. These data show that at equilibrium the concentration of the respective organic species in the two membrane layers were essentially equal to each other as were the concentrations in the two water zones.

The results show that the water in the bottom compartment had absorbed organics. At the end of the test the relative concentrations of the two organics were the same in both aqueous zones, demonstrating the transport of these organics through the geomembrane and the airspace to the water in the bottom compartment. The data also show that, for each of the two organics, the concentrations in the airspaces were similar, as were the concentrations in the two geomembrane specimens.

TABLE 9 -- Distribution of Organics in Three-Compartment
Test Apparatus Separated by Polyethylene Geomembranes [9]

Zone				Start of test		At end of test, 256 hours	
No.	Description	Volume, mL	Organic	Amount, mg	Concentration, mg/L	Amount, mg	Concentration, mg/L
1	Airspace above "cover" geomembrane[b]	806	TCE[a]	0	0	12.9	16.0
			Toluene	0	0	4.8	6.0
2	"Cover" poly- ethylene geo- membrane[b]	7.02	TCE	0	0	56.1	7,990
			Toluene	0	0	60.2	8,580
3	Airspace above test liquid	306	TCE	0	0	5.97	19.5
			Toluene	0	0	1.83	6.0
4	Test liquid	500	TCE	191	381	22.5	45.0
			Toluene	191	381	35.7	71.4
5	Barrier geo- membrane[b]	7.02	TCE	0	...	56.1	7,990
			Toluene	0	...	60.2	8,580
6	Airspace below geo- membrane	506	TCE	0	...	9.87	19.5
			Toluene	0	...	2.58	5.1
7	"Ground- water"	300	TCE	0	...	13.5	45.0
			Toluene	0	...	21.4	71.4
	Total	2,432	TCE	191	381	176.9	...
			Toluene	191	381	186.7	...
	Fraction ac- counted for, %		TCE	100		92.6	
			Toluene	100		97.7	

[a]Trichloroethylene.

[b]Linear low-density polyethylene (density 0.913; carbon black 2.5%).

Permeability of Geomembranes to Aqueous Solutions of Inorganic and Organic Species

The pouch test as described above can be used for assessing the simultaneous permeability of all components in a complex solution of both dilute organics and dilute inorganics. The configuration of the pouch is illustrated in Figure 3. It is necessary to track each component either by GC or GCMS for the organics and by EC or specific ion analysis for the inorganics, and the weight of the pouch for the water. The figure indicates the direction of migration of individual components from the pouch. If organics are present, it is necessary to seal the entire assembly in a closed container to avoid loss of organics and water.

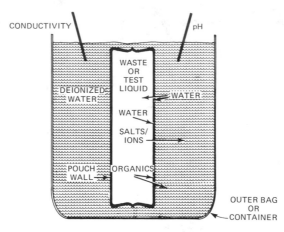

FIG. 3 -- Pouch assembly showing the movement of constituents
       during the pouch test. In the case illustrated by this
       drawing, the pouch is filled with an aqueous waste or
       test liquid and immersed in DI water. Arrows indicate
       the flow of specific constituents.

     The accuracy of the pouch test depends on preparation of durable,
leak-free pouches, the seams of which would not allow liquids to by-
pass the pouch wall and yield high transmission values. In work reported to date, the pouch test was restricted to thermoplastic geomembranes which could be heat-sealed or welded to make non-leaking
seams. The test should also apply to vulcanized geomembranes if
pouches can be fabricated to yield no leaks in the seams.

DISCUSSION

     The wide range of uses of geomembranes as barriers in many
different environments to many different permeating species requires
different test procedures to assess the effectiveness of a given
membrane for a given application. The permeating species range from
a single gas to highly complex mixtures such as those found in waste
liquids and leachates. In actual service it is important to know
actual transmission or migration of a species that would take place
under specific conditions and environments. Tests that would be
applicable to the measurement of permeability to different permeants
for various applications are summarized in Table 10.

     In the use of geomembranes in service as barriers to the trans-
mission of fluids, it is essential to recognize the difference between
geomembranes which are nonporous homogeneous materials and other liner
materials that are porous, such as soils and concretes. The transmission of permeating species through geomembranes without holes proceeds
by absorption of the species in the geomembrane and diffusion through
the geomembrane on a molecular basis. The driving force is chemical
potential across the geomembrane. A liquid permeates porous materials
in a condensed state that can carry the dissolved constituents, and

TABLE 10. Applicable Test Methods for Measuring
Permeability of Geomembranes to Various Permeants

Permeant	Example of permeant	Example of field application	Applicable test method	Reference
Single-component permeants				
Gas	$H_2, O_2,$ $N_2, CO_2,$ CH_4	Barriers, pipe and hose liners	D815, D1434-V, D1434-P	[2,3,4,5]
Water vapor	H_2O	Moisture vapor barriers	E96, D1653	[3,5]
Liquid water	H_2O	Liners for reservoirs, dams, and canals	Soil-type permeameter with hydraulic pressure	[6]
Organic vapor	Volatile solvents	Gasoline, secondary containment	D814, E96, F372	[4,5]
Organic liquid	Solvents	Containers, tank liners, secondary containment	D814, E96	[3]
Multi-component fluids				
Gases	CO_2/CH_4	Barriers, separation of gases	F372 CC, GCMS	[2,3]
Aqueous solutions of inorganics	Brines, combustion and incinerator ash leachates	Pond liners	Pouch, osmotic cell, ion analysis	[8]
Mixtures of organics	Spills, fuels	Liners for tanks and secondary containment	E96 with headspace GC	[3,4,5,8]
Aqueous solutions of organics	Waste liquids and leachates	Liners for ponds and waste disposal	Pouch, multi-component cell	[9]
Complex aqueous solutions of organics and inorganic species	Waste liquids and leachates	Liners for waste disposal	Pouch, multi-component cell, osmotic cell, head-space GC, GCMS	[5]

the driving force for such permeation is hydraulic pressure. Due to the selective nature of geomembranes, the permeation of the dissolved constituents in liquids can vary greatly, i.e., some components of a mixture can permeate more readily than others. With respect to the inorganic aqueous salt solution, the geomembranes are semipermeable, i.e., the water can be transmitted through the geomembranes, but the ions are not transmitted. Thus, the water that is transmitted through a hole-free geomembrane does not carry inorganics. The direction of permeation is determined by the chemical potential across the geomembrane. The wastewater on the upstream side is at a lower potential than the less contaminated water on the downstream side which can permeate into the wastewater by osmosis.

Although inorganic salts do not permeate geomembranes, some organic species do. The rate of permeation through a geomembrane depends on the solubility of the organic in the geomembrane and the diffusibility of the organic in the geomembrane as driven by the chemical potential gradient. Factors that can affect the diffusion of an organic within a geomembrane include:

1. The microstructure of the polymer, e.g., percent crystallinity.
2. The glass transition temperature of the polymer.
3. The other constituents in the geomembrane compound such as plasticizers and fillers.
4. The flexibility of the polymer chains.
5. The size and shape of the diffusing molecules.
6. The temperature at which diffusion is taking place.

The movement through a hole-free geomembrane of mobile species that would be encountered in service would be affected by many factors, such as:

1. The composition of the geomembrane with respect to the polymer and to the compound.
2. The thickness of the geomembrane.
3. The service temperature.
4. The temperature gradient across the geomembrane in service.
5. The chemical potential across the geomembrane, which includes pressure and concentration gradient.
6. The composition of the fluid and the mobile constituents.
7. The solubility of various components in the particular geomembrane.
8. The ion concentration of the liquid.
9. Ability of the species to move away from the surface on the downstream side.

Because of the great number of variables, it is important to perform permeability tests of a geomembrane under conditions that simulate as closely as possible the actual environmental conditions in which the geomembrane will be in service.

REFERENCES

[1] Daniel, D. E., Shackelford, C. D., and Liao, W. P., "Transport of
     Inorganic Compounds Through Compacted Clay Soil," Proceedings of
     the Fourteenth Annual Solid Waste Research Symposium on Land Dis-
     posal, Remedial Action, Incineration and Treatment of Hazardous
     Waste, EPA/600/9-88/021, U.S. Environmental Protection Agency,
     Cincinnati, OH, 1988, pp. 145-166.

2.   Yasuda, H., Clark, H. G., and Stannett, V. "Permeability,"
     in Encyclopedia of Polymer Science and Technology, Vol. 9,
     Interscience, NY, 1968, pp. 794-807.

3.   ASTM. Annual Book of ASTM Standards. Issued annually in several
     parts. American Society for Testing and Materials, Philadelphia,
     PA:

          D471-79.   "Test Method for Rubber Property--Effect of
                     Liquids."

          D814-86.   "Test Method for Rubber Property--Vapor Trans-
                     mission of Volatile Liquids."

          D815-81.   "Standard Method for Testing Coated Fabrics--
                     Hydrogen Permeability."

          D1434-84. "Test Method for Determining Gas Permeability
                     Characteristics of Plastic Film and Sheeting to
                     Gases."

          D1653-85. "Standard Test Methods for Water Vapor Perme-
                     ability of Organic Coating Films."

          E96-80.   "Test Methods for Water Vapor Transmission of
                     Materials."

          F372-73 (Reapproved 1984). "Standard Test Method for Water
                     Vapor Transmission rate of Flexible Barrier
                     Materials Using an Infrared Detection Technique."

          F739-85.   "Standard Test Method for Resistance of Protec-
                     tive Clothing Materials to Permeation by Liquids
                     or Gases."

4.   Haxo, H. E., Miedema, J. A., and Nelson, N. A. "Permeability of
     Polymeric Membrane Lining Materials for Waste Management Facil-
     ities," Elastomerics, Vol. 117, No. 5, 1985, pp. 29-26,66.

5.   Matrecon, Inc., "Lining of Waste Containment and Other Impoundment
     Facilities," 2nd Revised Edition, EPA/600/2-88/052, U.S. Environ-
     mental Protection Agency, Cincinnati, OH, 1988, 991 pp.

6.   Giroud, J. P., "Impermeability: The Myth and a Rational Apporach,"
     Proceedings of the International Conference on Geomembranes,
     Denver, CO, Volume I, Sponsored by Industrial Fabrics Association
     International, St. Paul, MN, 1984, pp. 157-162.

7.   Giroud, J. P, and Bonaparte, R., "Leakage through Liners Con-
     structed with Geomembranes--Part I.   Geomembrane Liners," Geo-
     textiles and Geomembranes, Vol. 8., 1989, pp. 27-67.

8.   August, H., and Tatzky, R., "Permeabilities of Commercially Avail-
     able Polymeric Liners for Hazardous Landfill Leachate Organic
     Constituents," Proceedings of the International Conference on Geo-
     membranes," Denver, CO, Vol. 1, Industrial Fabrics Association
     International, St. Paul, MN, 1984, pp. 163-168.

9.   Haxo, H. E., and Lahey, T. P., "Transport of Dissolved Organics
     from Dilute Aqueous Solutions Through Flexible Membrane Liners,"
     Hazardous Waste & Hazardous Materials, Vol. 5, No. 4, Mary Ann
     Liebert, Inc., publishers, NY, 1988, pp. 275-294.

Yick H. Halse, Arthur E.Lord, Jr. and Robert M. Koerner

DUCTILE-TO-BRITTLE TRANSITION TIME IN POLYETHYLENE GEOMEMBRANE SHEET

REFERENCE : Halse, Y.H., Lord, A.E. Jr., Koerner, R.M.,
"Ductile-To-Brittle Transition Time in Polyethylene
Geomembrane Sheet" Geosynthetic Testing for Waste
Containment Applications, ASTM STP 1081, Robert M.
Koerner, editor, American Society for Testing and
Materials, Philadelphia, 1990.

ABSTRACT: The susceptibility of polyethylene (PE) geomembranes
to stress cracking was evaluated in the laboratory using an
accelerated notched constant load testing (NCLT) method. The
test specimens were subjected to various stress levels which
ranged from 25% to 70% of the yield stress at 50°C. The
ductile-to-brittle failure curves of five different
geomembranes were obtained by plotting the logarithm of percent
yield stress against the logarithm of average failure time.
The stress cracking resistance (SCR) is quantified as the value
of the ductile-to-brittle transition time. The results show
that SCR of these geomembrane sheets vary over a wide range
from a minimum of 4 hours to over 600 hours.

KEYWORDS: geomembrane, polyethylene, ductile, brittle, stress
cracking resistance, crystallinity, amorphous, lamellae, tie
molecules, ductile-brittle transition, fracture, fiber, hackle

INTRODUCTION

   Polyethylene (PE) is a widely used resin in making geomembrane
sheets. This is due to its high chemical resistance which increases
with increasing crystallinity of the material. However, concerns
have been raised in the use of this highly crystalline material (both
sheet and seams) regarding its stress cracking resistance (SCR)
behavior [1-3].

   The widely accepted method for evaluating SCR of geomembrane sheet
is the ASTM D-1693[4] bent strip test method. This is a constant
strain test in which the specimens are confined into a fixed

   Drs. Halse, Lord and Koerner are Research Assistant Professor,
Professor and Director respectively, at the Geosynthetic Research
Institute, Drexel University, Philadelphia, PA 19104.

configuration for the duration of the experiment. The initially induced stress is high, but it decreases with time due to polymer relaxation. Therefore, the test may not be meaningful after a certain length of time. Furthermore, with recent advances in PE polymerization and stabilization, the test may not be sensitive enough to distinguish between subtle changes in resin.

This paper proposes a notched constant load test (NCLT) method for evaluating geomembrane sheet material. The emphasis of the method is on quantitative comparison of SCR between different geomembrane sheets.

DUCTILE-TO-BRITTLE TRANSITION PHENOMENA

It is commonly known that PE fails in a ductile manner under relatively high loads. Depending on the test configuration, the material can elongate over 1000% of its original length becoming essentially a thin film with respect to its original thickness. However, the same material can also fail in brittle manner characterized by relatively little deformation in the failure area. Such brittle failure is often associated with long time frames and low stress conditions. It is commonly referred to as "stress cracking". According to ASTM D-883[5] stress cracking is defined as "an external or internal rupture in a plastic caused by tensile stress less than its short-term mechanical strength".

Based on the chain fold model[6,7], Lustiger[8] has explained ductile and brittle failures from the molecular point of view. The basic crystal structure in PE is an integrated combination of crystalline lamellae and amorphous phases. The crystal phase is formed by molecular chain folding, as illustrated in Figure 1. The space between lamellae is the amorphous phase. Molecular chains linking two adjacent lamellae together are called "tie molecules" which probably contribute the most to transferring load between lamellae.

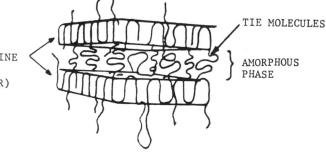

Fig. 1 -- A simplified schematic diagram of PE microstructure

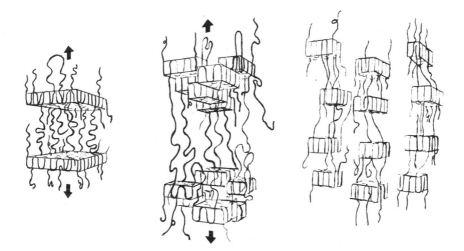

Fig. 2 -- Steps in the ductile deformation of polyethlene

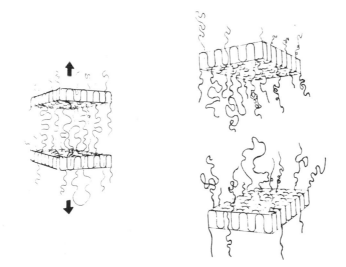

Fig. 3 -- Steps in the slow crack growth of polyethylene

In ductile failure, the tie molecules stretch to their maximum as shown in Figure 2. The cohesive force within the lamella is weaker than the strength of the tie molecules which results in breaking down of lamella into numerous small units called "mosaic blocks"[7]. This forms a series of fibers creating the thin film commonly seen in index testing of HDPE specimens, e.g. ASTM D-638[9].

In brittle failure, the applied stress is much lower than the stress required to cause ductile deformation as discussed previously. The tie molecules are under tension but are not fully stretched. Over a long period of time, the tie molecules can be broken due to relaxation and distanglement or chain scission, as shown in Figure 3. This results in failure surfaces having very little sign of plastic deformation, i.e. it is characterized as a brittle failure.

An explanation of the transition from ductile-to-brittle failure requires a knowledge of the stress level, the stress concentration factor, the temperature, and the surrounding environment. It is obviously strongly affected by the characteristics of the base resin, among which items are molecular weight distribution, type of comonomer, and the density of branching[10-12]. These factors are controlled by the polymerization and blending process of the resins. Additionally, the manufacturing process of the geomembrane sheet may also have some effects. Therefore, the test should use samples in their final manufactured form and not reconstituted into idealized shapes. Beside all of the above considerations, the ductile-to-brittle transition is also very sensitive to the aging and degradation of the polymer. Choi and Broutman[13] have found that there is a 90% reduction in the failure time after 250 hours of UV exposure, and 70% to 80% reduction after 50 hours. This strongly suggests that UV degradation can significantly promote brittle failure.

Whatever the situation, brittle failures are simply not desirable. In this regard, the longer the transition time between ductile and brittle failures, the better is the material in its stress cracking performance. It is vital to know when such transition occurs and the conditions in which it is occurring.

DATA EXTRAPOLATION METHODS

Ideally, the performance of the material should be evaluated by testing at its design stress level under an environment closely simulating the field situation. However, the typical field situation for geomembrane liners is probably between 10° to 30°C and generally immersed. The testing time under such condition can be very long. Therefore, acceleration procedures are often utilized which include raising the temperature and immersion in a surface active wetting agent. After a reasonable test time, data extrapolation can be applied to predict the ductile-to-brittle transition at ambient temperature.

The various equations used to predict long term performance of PE gas pipe have been reviewed in detail by Lord and Halse[14]. The method which is widely used in the natural gas plastic pipe industry is the Rate Process Method (RPM)[15,16]. It is used to predict a failure curve at some temperatures other than those tested. This

method is vased on an absolute reaction rate theory as developed by
Tobolsky and Eyring[17] for viscoelastic phenomenon. Coleman [18] has
applied it to explain the failure of polymeric fibers. The
relationship between failure time and stress is expressed in the form
as follows:

$$\log t_f = A_0 T^{-1} + A_1 T^{-1} \sigma \qquad\qquad (1)$$

where:
t_f = time to failure
T = temperature
σ = Tensile stress on the fiber
A_0 and A_1 are constants

Bragaw[15] has revised the above model on polymeric fibers and found
three additional equations which yield reasonable correlation to the
failure data of HDPE pipe. These three equations are as follows:

$$\log t_f = A_0 + A_1 T^{-1} + (A_2 T^{-1})P \qquad\qquad (2)$$
$$\log t_f = A_0 + A_1 T^{-1} + A_2 \log P \qquad\qquad (3)$$
$$\log t_f = A_0 + A_1 T^{-1} + (A_2 T^{-1}) \log P \qquad\qquad (4)$$

where:
P = internal pipe pressure proportional to the hoop stress in the
     pipe
A_0, A_1, and A_2 are constants

The application of RPM requires a minimum of two experimental
failure curves at different elevated temperatures generally above
40°C. The equation which yields the best correlation to these curves
is then used in the prediction procedure for a response curve at a
field related temperature e.g., 10°C to 25°C. (For data plotted in a
log-normal scale, only the equation (2) will be considered, whereas in
log-log scale, both equations (3) and (4) should be considered.) Two
separate extrapolations are required, one for the ductile response and
one for the brittle response. Three representative points are chosen
on the ductile regions of the two experimental curves. One curve will
be selected for two points, and the other, the remaining point. This
data is substituted into the chosen equation, i.e., Equation 2, 3, or
4 to obtain the three constants by solving the simultaneous equations.
The constant values are then substituted back to the chosen equation
to form the prediction equation for the ductile response of the curve
at the desired temperature. The process is now repeated for the
predicted brittle response curve at the same desired temperature.

EXPERIMENTAL MATERIALS

Five commercially available PE geomembrane sheet materials were
used in this study. They were made from five different resins and
three different manufacturing methods. Materials S-1, S-4 and S-5
geomembrane sheets were manufactured by the same process but used
different PE resins. Material S-1 is ten years old and had been
constantly exposed to outdoor environment in the southeast region of
the US. Materials S-2 and S-4 are one year old and had also been

exposed to ambient environment in the northeast and southeast region of the U.S. respectively. Materials S-3 and S-5 are newly manufactured sheets and have never been exposed. Some of the properties of each material are tabulated in Table 1.

Table 1 -- Properties of PE geomembranes used in this study.

Mate-rial	Age	Thick-ness	Density*	Melt Index	FRR+	Yield Stress at 25°C	Bent Strip Results** (no.failures/
	(year)	(mm)	(g/cc)	(g/10 min)	(P/E)	(Mpa)	hours of test)
S-1	10	2.4	0.954	0.19	3.0	23	9/700
S-2	1	2.0	0.950	0.57	2.5	20	6/700
S-3	New	1.9	0.948	0.17	4.9	20	0/2000
S-4	1	2.2	0.955	0.28	2.4	21	5/2000
S-5	New	2.2	0.950	0.18	5.0	20	0/2000

*   The density values represent the density of the final geomembrane compound which contains 2 to 2.5% of carbon black and an unknown processing package.
+   FRR = Flow Rate Ratio after ASTM D-1238 [19].
** Total of ten specimens in each test. Test conditions are 50°C and 10% Igepal solution using ASTM D-1693.

EXPERIMENTAL PROCEDURES

Specimen Preparation

A dumbbell shaped specimen was cut out from the geomembrane sheet using a ASTM D-1822 die, (see Figure 4). The orientation of the length of the specimen was parallel to the cross machine direction. A notch was introduced at the center of the neck on one of the surfaces as can be seen in Figure 5. Thus the notch is in the machine direction of the manufactured sheet. The depth of the notch is 20% (±0.5%) of the thickness of the specimen and it was made with a single edge razor blade. The blade was mounted in the upper jaw grip of a compression testing machine. The notching was performed at a rate of 0.13 mm/min (0.005 in/min). A new blade was used for notching of every ten test specimens.

Test Procedure

The notched test specimens were fitted with grommets in their end tabs and placed in the test apparatus. The design of this apparatus is based on that used in ASTM D-2552[20]. Its front and side views are shown in Figure 6. The device is capable of testing up to 20 specimens simultaneously. The specimens were subjected to a predetermined constant stress while immersed in a 10% Igepal (CO-630) solution with tap water at 50°C. The failure time of the specimens was automatically recorded to the nearest 0.1 hour. Failure was visually evidenced by either a ductile elongation or a brittle cracking.

The stress levels used in the tests were ranged from 25% to 70% of

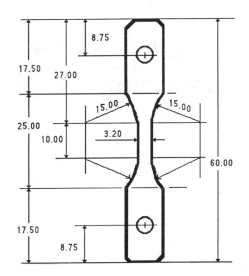

Fig. 4 -- Dimensions of test specimen

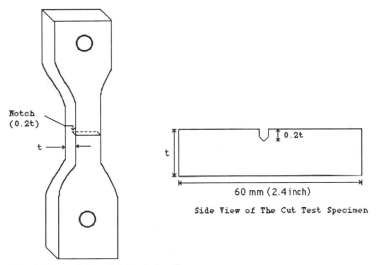

Front View of the Cut Test Specimen

Side View of The Cut Test Specimen

Fig. 5 -- Schematic diagram of a notched specimen

the yield stress at 50°C. The maximum stress increments were never greater than 5%. Three specimens were tested at each stress level and the average value was used for curve plotting.

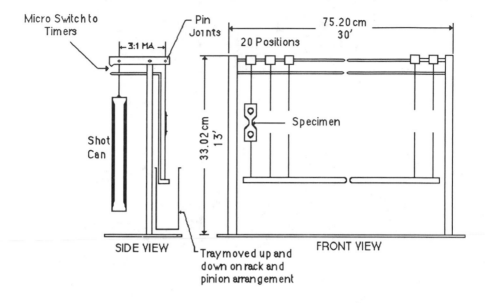

Fig. 6 -- Schematic diagram of notched constant
load test device.

RESULTS AND DISCUSSIONS

This section of the paper presents the resulting test data of the five different geomembrane materials. As a supplement to this data, fracture morphology was also examined to illustrate the differences between ductile and brittle failures which will be shown later in this paper.

Ductile-to-Brittle Transition

The experimental data is analyzed by plotting the logarithm of applied stress (which was calculated as the percentage of the yield stress at 50°C) against the logarithm of failure time of each of the specimens. The resulting ductile-to-brittle curve can be visually separated into two linear regions, (see Figure 7). At high stresses, the slope of the line is relatively shallow. In this region, the specimens exhibit ductile failures. At low stresses the slope of the line becomes much steeper. The specimens corresponding to this region failed in a brittle mode. The time corresponding to the point at which slope changes is called the ductile-to-brittle transition time, and designated as "T_t".

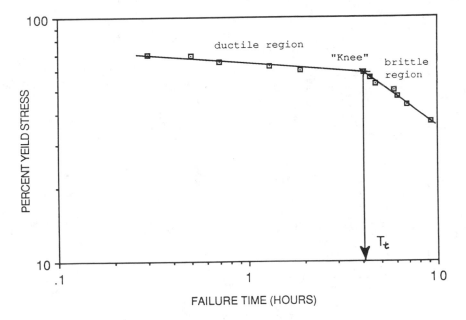

Fig. 7 -- Ductile-brittle curve of PE geomembrane sheet,
        showing "knee" shape transition, at 50°C and 10% Igepal

    Figure 8 displays the ductile-to-brittle curves of four of the
different geomembrane sheet materials. All exhibit a distinct
transition, showing a pronounced "knee" in the curve. The
ductile-to-brittle transition of each is very pronounced, particularly
materials S-1 and S-2. To be noted is that the transition times of
these two sheets are only a few hours long. In contrast, materials
S-3 and S-5 show a smaller change in their slopes as the failure mode
changes. Their transition times are of the order of a few hundred
hours. The results suggest that materials S-3 and S-5 have much
better SCR behavior than materials S-1 and S-2 under these test
conditions. The slopes of the ductile and brittle regions of these
four geomembranes along with their transition times are listed in
Table 2.

    The response curve of material S-4 is slightly different than
those shown in Figure 8. It is given in Figure 9. Instead of a
"knee" shape transition, an overshoot region (called a "nose")
appeared at the transition between ductile and brittle behavior. The
ductile-to-brittle transition time corresponds to the onset of brittle
region, indicated as Point "A" in Figure 9. The slopes of the ductile
and brittle regions of material S-4 along with its transition time are
also listed in Table 2.

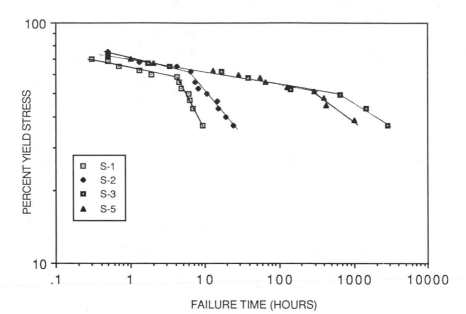

Fig. 8 -- Ductile-brittle curves of four different geomembrane
sheets, S-1, S-2, S-3, and S-5, at 50°C and 10% Igepal

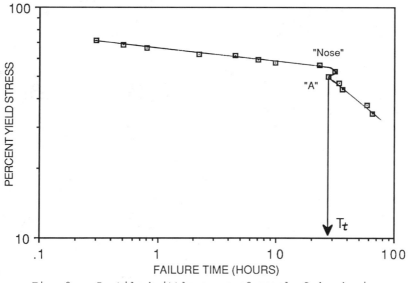

Fig. 9 -- Ductile-brittle curve of sample S-4, showing
a "Nose" shape transition, at 50°C and 10% Igepal

Table 2 -- Summary of ductile-to-brittle behaviours of five different geomembrane sheets evaluated in this study.

Materials	Shape of Transition	Ductile Region Slope (%/hour)	Brittle Region Slope (%/hour)	Transition Time (hour)
S-1	Knee	-12.7	-1.9	4
S-2	Knee	-22.2	-2.6	6
S-3	Knee	-18.1	-4.4	600
S-4	Nose	-15.4	-2.1	28
S-5	Knee	-16.8	-5.0	300

The data in Table 2 reflects that there are large variations between these five geomembrane materials. The short transition time of material S-1 is probably affected by both the quality of the resin and outdoor exposure, since the geomembrane was made and installed ten years ago. For the two newly manufactured geomembranes, S-3 and S-5, the transition times are substantially longer than the others. This may be partially due to improvements in the properties of resins regarding SCR. However, the environmental exposure effects may also be a factor, since both materials have not been exposed to direct sunlight.

It is somewhat disconcerting, however, that there is no clear correlation between the transition times and properties of the materials listed in Table 1, except for the bent strip SCR test results. For materials S-3 and S-5, which have transition times of a few hundred hours, no cracks, were observed within the 2000 hour bent strip testing times. On the other hand, the poor stress cracking performance of materials S-1 and S-2 was also reflected in the results of bent strip tests; high percentages of specimens were cracked after only 700 hours. The bent strip test results of material S-4 falls somewhere in between the results of the other materials.

The consistency between the proposed NCLT test and the bent strip test indicates that both methods are valuable to evaluate the SCR of HDPE geomembrane sheets, particularly for the poor SCR materials. However, for materials with transition times greater than 100 hours, bent strip testing is not capable to distinguish between materials. In addition, the bent strip test is a qualitative test, whereas NCLT is a quantitative one. Furthermore, the ductile-to-brittle curve could be used to monitor the quality of the seams by the "seam factor" (see reference 13) and to extrapolate the performance of the geomembrane to site specific temperatures.

Fracture Morphology

The fracture morphology of failure specimens from regions of ductile, brittle and transition zone region were examined using a scanning electron microscope (SEM). The micrograph of a failed specimen in the ductile region can be seen in Figure 10, which shows an uniform pull-out of a group of fibers. For brittle failure, the fracture morphology can be seen in Figure 11. It is dominated by

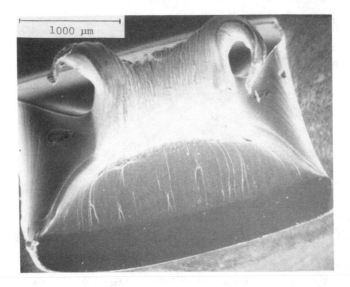

Fig. 10 -- Fracture surface of a specimen failed
in ductile region

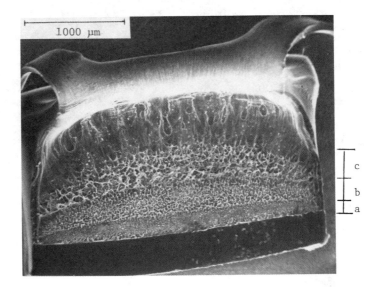

Fig. 11 -- Fracture surface of a specimen failed
in brittle region

short fiber structure with a small section of hackle structure. (See reference [21] for description of these terms). Three regions of Figure 11, a, b, and c, can be seen in the fiber section, and they are distinguished by the size of the fibers. The size of the fibers increase as stress acting on the crack tip increases due to reduction in cross sectional area. The divisions in the fiber section also suggests that crack arrest took place, hence crack growth probably is not linear with time. The fracture surface of a specimen which failed in the transition region can be seen in Figure 12. The dominate morphology is hackle with a small section of fiber structure near the initial failure area.

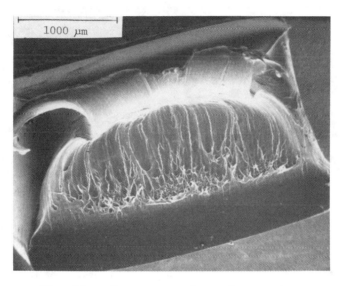

Fig. 12 -- Fracture surface of a specimen
failed in transition region.

SUMMARY

The susceptibility of polyethylene geomembrane sheets to stress cracking is evaluated using a NCLT method. The performance of the materials was indicated by the ductile-to-brittle transition time value. Five different types of geomembranes were studied, and their transition times vary significantly. Materials S-1, S-2, and S-4 exhibit a rather short transition time, from 4 to 28 hours, whereas materials S-3 and S-5 show a relatively long transition time, from 300 to 600 hours. This indicates that the latter two geomembranes have much better SCR behavior than the other three. The variation in stress cracking performance is probably caused by a number of factors, among which are:

• The amount of outdoor degradation, for example, UV and thermal degradation can induce chain scission and cross linking in PE, hence reducing SCR.

- The age of sheet, in which reducing SCR aging can increase the density of the geomembrane by the process of secondary crystallization.
- The method of manufacturing and handling the process may induce residual stresses in the final product, and assist the propagation of stress cracks.

As such, we feel that the ductile-to-brittle transition time can not only evaluate the susceptibility of different geomembrane sheets regarding to SCR, it also can be used to monitor changes in the sheet.

There is a correlation between ductile-to-brittle transition time and bent strip test results, particularly for the relative poor SCR materials. However, the bent strip test cannot distinguish materials which have 300 hours difference in their transition times. This indicates that the constant tensile stress test is a much more sensitive test than the bent strip test. In addition, it is a quantitative test. The data could also be used via RPM to extrapolate the ductile-to-brittle transition time of the geomembranes at the site specific ambient temperature.

ACKNOWLEDGEMENTS

This project is funded by the U.S. Environmental Protection Agency under Project No. CR-815692-01. Our sincere appreciation is extended to the Agency and to our Project Officer, David A. Carson.

REFERENCES

[1]   Halse, Y. H., Koerner, R. M., and Lord, A. E. Jr., "Laboratory Evaluation of Stress Cracking in HDPE Geomembrane Seams", Aging and Durability of Geosynthetics, Koerner, R. M., Ed., Elsevier Applied Science, 1989.

[2]   Peggs, I. A., and Carlson, D. S., "Stress Cracking of Polyethylene Geomembranes : Field Experience", Aging and Durability of Geosynthetics, Koerner, R. M., Ed., Elsevier Applied Science, 1989.

[3]   Fisher, G. E., "Controlling Thermal damage in Flexible Membrane Liners", Geotech. Fabrics Report, Mar./Apr., 1989, IFAI, pp. 39.

[4]   ASTM D-1693, "Environmental Stress Cracking of Ethylene Plastics", ASTM, Vol.08.02, 1988.

[5]   ASTM D-883, "Standard Definitions of Terms Relating to Plastics", ASTM, Vol.08.01, 1988.

[6]   DiMarzio, E. A., and Guttmann, C. M., "Three Statistical Mechanical Arguments That Favor Chain Folding in Polymer Systems of Lamellar Morphology", Polymer, 1980, 21, pp. 733.

[7]   Peterlin, A., "Morphology and Fracture of Draw Crystalline Polymers", Journal Macromol. Sci., Phys. 1973, B8, pp. 83.

[8]   Lustiger, A., "The Molecular Mechanism of Slow Crack Growth in Polyethylene", Eighth Plastic Fuel Gas Pipe Symposium", 1983, pp. 54.

[9]   ASTM D-638, "Tensile Properties of Plastics", ASTM, Vol.08.01, 1988.

[10] Lu, X, and Brown, N., "Effect of thermal History on the Initiation of Slow Crack Growth in Linear Polyethylene", Polymer, 26, 1985

[11] Raske, R. T., "Analysis and Application of the Constant Tensile Load Test for Polyethylene Gas Pipe Materials", Tenth Plastic Fuel Gas Pipe Symposium, 1987, pp. 102.

[12] Bubeck, R. A., and Baker, H. M., "The influence of Branch Length on the Deformation and Microstructure of Polyethylene", Polymer, 23, 1982

[13] Choi, S. W. and Broutman L. J., "Ductile-Brittle Transitions for Polyethylene Pipe Grade Resins" Eleven Plastic Fuel Gas Pipe Symposium, 1989, pp. 296.

[14] Lord, A. E., Jr., and Halse, Y., "Polymer Durability - The Materials Aspects", Aging and Durability of Geosynthetics, Koerner, R. M., Ed., Elsevier Applied Science, 1989.

[15] Bragaw C. G., "Service Rating of Polyethylene Piping Systems by The Rate Process Method", Eighth Plastic Fuel Gas Pipe Symposium, 1983, pp. 40.

[16] Twura, S. and Nishio, N., "New Testing Method for Predicting Service Life of Polyethylene" Eighth Plastic Fuel Gas Pipe Symposium, 1983, pp. 34.

[17] Toabolsky, A. and Eyring, H., "Mechanical Properties of Polymeric Materials," Journal of Chem. Phys., Vol.11, 1943, pp. 125-134.

[18] Coleman, B.D., "Application of the Theory of Absolute Reaction Rates to the Creep Failure of Polymeric filaments," Journal Polymer Science, Vol. 20, 1956, pp. 447-455.

[19] ASTM D-1238, "Flow Rates of Thermoplastic by Extrusion Plastometer," ASTM, Vol. 08.01, 1988.

[20] ASTM D-2552, "Test Method for Environmental Stress Rupture of Type III Polyethylenes Under Constant Tensile Load", ASTM, Vol.08.02, 1987.

[21] Halse, Y.H., Lord, A.E. Jr., and Keorner, R.M., "Stress Cracking Morphology of HDPE Geomembrane Seams", Microstructure and the Performance of Geosynthetics, ASTM STP ___, Ian Poggs, editor, ASTM, Philadelphia, 1989.

Greg A. Whyatt, Rick K. Farnsworth[a]

THE HIGH pH CHEMICAL AND RADIATION COMPATIBILITY OF VARIOUS LINER MATERIALS

REFERENCE: Whyatt, G. A., and Farnsworth, R. K. "The High pH Chemical and Radiation Compatibility of Various Liner Materials," Geosynthetic Testing for Waste Containment Applications, ASTM STP 1081, Robert M. Koerner, editor, American Society for Testing and Materials, Philadelphia, 1990.

ABSTRACT: A flexible membrane liner has been proposed to line a concrete vault in which liquid low-level radioactive waste will be solidified. High-density polyethylene (HDPE) and polypropylene liners were tested at the Pacific Northwest Laboratory[b] in an EPA method 9090 format to determine their chemical compatibility with the waste. Radiation effects were also investigated. The liners were immersed in a highly caustic (pH>14), primarily inorganic solution at 90°C. The liners were subjected to radiation doses up to 38.9 Mrad, which was the expected dose the liner would receive over a 30-year life inside the vault. Recent changes have placed the liner outside the vault.

The acceptance criteria for judging the compatibility of the liner with radiation should be different than those used for judging chemical compatibility. The radiation damage over the life of the liner can be simulated in a short-term test. Both HDPE and polypropylene liners were judged to be acceptable from a chemical and radiation standpoint when placed outside of the vault, while several other liners were not compatible. Radiation did not have a significant effect on chemical degradation rates.

KEYWORDS: high-density polyethylene, HDPE, polyproplyene, geomembrane, liner, radiation, compatibility.

(a) G. A. Whyatt and R. K. Farnsworth are researchers for Battelle Northwest, Battelle Boulevard, Richland, Washington 99352.
(b) Pacific Northwest Laboratory is operated by Battelle Memorial Institute for the U.S. Department of Energy under Contract DE-AC06-76RLO 1830. The study reported here was funded by Westinghouse Hanford Company in support of the Grout Disposal Program.

INTRODUCTION

At the Hanford Site near Richland, Washington, plans currently exist for disposing of low-level liquid waste in a grout form. Grout is a mixture of liquid waste and grout formers, including fly ash, blast furnace slag, and Portland cement. The liquid waste will be mixed with the grout formers and the resulting slurry will be pumped to concrete vaults, where it will harden into large grout masses. The waste stream considered here [double-shell slurry feed (DSSF)] is a very high pH (>14) solution containing high concentrations of inorganic compounds and very small concentrations of organic compounds. Because the waste contains hazardous constituents, its disposal is governed by the Resource Conservation and Recovery Act (RCRA) (Public Law 94-580) [1]. Therefore, the liners to be used in the vault must be shown to be compatible with the DSSF grout slurry and any leachate that might be generated from the DSSF grout.

Exothermic hydration reactions occurring after the grout is poured in the vault, combined with radiolytic heat, may produce temperatures as high as 90°C for extended periods. The liners will receive a radiation dose of 38.9 Mrad over 30 years (after this work was completed, a design change placed the liner outside of the vault, thereby reducing the 30-year dose to approximately 14 rad and potentially reducing temperatures). To determine the compatibility of the liners with the waste, the testing was performed in accordance with methods specified by the Environmental Protection Agency (EPA), Method 9090 Compatibility Test for Wastes and Membrane Liners (EPA 9090 test) [2].

EXPERIMENTAL PROCEDURE

Immersion Procedure

The EPA 9090 test method calls for immersion of liner samples in leachate. The disposal scenario considered here is unique since the grout is pumped into the vault as a liquid and then solidifies. Leaching of the grout is not expected due to barriers over the system, which limit recharge and divert advecting water away from the grout. Therefore, no attempt was made to simulate a leachate from the grout. Instead, the liners were immersed directly in a simulated, nonradioactive DSSF waste. The simulated waste was produced by adding various chemicals to match the concentrations in the actual waste. All radionuclides were omitted from the simulated waste. Since the tests were conducted, the estimate of the waste composition has been modified. Table 1 compares the waste composition used in these tests, measured using inductively coupled plasma (ICP) spectroscopy, to the current estimate of the waste composition. The immersion temperature was 90°C, which is the maximum temperature expected in the grout. With the current design, the immersion in simulated DSSF represented a conservatively severe condition for the following reasons:

TABLE 1 -- Comparison of immersion solution to actual waste composition

Component	Measured Molarity	Actual Expected Waste Molarity [3]
Ag	1.8×10^{-3}	5.18×10^{-5}
Al	0.47	0.578
Ba	4.7×10^{-3}	4.35×10^{-5}
Ca	5.5×10^{-3}	1.17×10^{-3}
Cd	6.55×10^{-5}	1.39×10^{-4}
Cl	8.0×10^{-2}	9.9×10^{-2}
Cr	3.0×10^{-2}	7.46×10^{-3}
Cu	5.25×10^{-5}	7.16×10^{-5}
F	$<4.5 \times 10^{-2}$	1.98×10^{-2}
Fe	2.73×10^{-2}	3.49×10^{-4}
Hg	1.3×10^{-5}	2.33×10^{-5}
K	0.21	0.233
Mn	4.8×10^{-2}	1.70×10^{-4}
Mo	3.1×10^{-4}	3.52×10^{-4}
Na	6.0	5.65
NO_2	1.9	0.961
NO_3	1.5	1.64
Ni	5.10×10^{-4}	4.65×10^{-4}
P	4.3×10^{-2}	5.75×10^{-2}
Pb	7.5×10^{-6}	3.95×10^{-4}
Se	3.4×10^{-5}	3.62×10^{-4}
SO_4	0.11	2.03×10^{-2}
Zn	1.6×10^{-2}	1.79×10^{-4}
EDTA	8.3×10^{-3} (a)	8.10×10^{-3}
Citrate	8.4×10^{-3} (a)	7.00×10^{-3}

(a) Represents amount added in simulant formulation, not analysis value.

- Reactions with the solid grout formers decrease the pH of the waste. Also, leachate would be diluted and therefore would have less effect.

- During a pilot-scale test, all separated liquid (if any existed) was absorbed by the grout in less than 24 hours. Therefore, any leak would only be of short duration.

- The disposal system incorporates barriers that divert advecting water away from the grout. The area of the system draining to the liner is gravel and should drain relatively quickly compared to a landfill after closure. Therefore, the duration and quantity of leachate is less than expected in a landfill after closure.

The liners tested included the following:

1) High-density polyethylene (HDPE) liner - 1.52 mm (60 mil)

2) Polypropylene liner - 4.32 mm (170 mil) polypropylene
3) Nonreinforced Hytrel polyester liner - 1.02 mm (40 mil)
4) Polyurethane liner - 1.52 mm (60 mil)
5) Ethylene interpolymer alloy (EIA) coated polyester fabric - .89 mm (35 mil)

Per the standard EPA 9090 test format, samples were tested before immersion in the waste and after immersions of 30, 60, 90, and 120 days. A detailed description of the procedures used for measuring various properties is provided in Farnsworth and Hymas (1989) [4].

Irradiation Procedure

Additional liner samples were tested to determine radiation compatibility. Samples were first immersed in DSSF for 30 days at 90°C and then placed in a gamma irradiation pit and irradiated to a dose of 0.6, 3.6, 16.1, or 38.9 Mrad while immersed in DSSF. Those doses represent the expected dose for material inside the grout vault over periods of 120 days, 2 years, 10 years, and 30 years based on the radionuclide inventory in the waste [3]. Samples were tested at each exposure level after removal from the irradiation pit. In addition, samples at the 0.6 Mrad and 38.9 Mrad levels were tested after an additional 90 days of immersion to determine if radiation exposure increases the susceptibility of the liner to chemical attack from the waste.

Selection of a 30-year dose as the maximum exposure value was based on guidelines from RCRA, which specify that disposal facilities must be monitored for 30 years after closure [1]. The purpose of a 30-year post-closure period is to permit drainage of a landfill that has been exposed to precipitation. The grout disposal system is not expected to generate leachate over a long period of time and therefore may not require a leachate collection system to be functional for a 30-year life.

The rate and type of radiation exposure in the tests differed from the low rate of alpha, beta, and gamma radiation that would be expected in actual service. However, many studies have shown that radiation damage is only dependent on total dose, not dose rate or type of radiation [5,6,7].

ACCEPTANCE CRITERIA

The chemical suitability of a liner for a particular service can be judged from the results of compatibility tests. Compatibility tests attempt to predict the long-term (30 year) compatibility from short-term (120 day) changes in properties. Compatibility is judged based on the magnitude of changes in liner properties and whether the changes stabilize over the 120 days of the test. Except in the case where a liner is clearly not compatible, criteria are needed to help determine whether the liner is chemically suited for a particular service.

For the HDPE liner, the acceptance criteria suggested by the National Sanitation Foundation (NSF) [8] were used. The criteria includes a reference to the minimum as-received values for HDPE, as defined by NSF Standard 54 [9]. The published NSF criteria for HDPE liners are as follows:

- stability of weight change and mechanical properties with time

- a stabilized weight gain of not more than 3%

- a breaking factor (or strength) at least 80% of the initial value, and equal to or greater than the minimum as-received value in the material properties table of NSF Standard 54

- percent elongation at break at least 80% of the initial value, and equal to or greater than the minimum as-received value in the material property table of NSF Standard 54

- yield strength at least 80% of the initial value and equal to or greater than the minimum as-received value in the material property table of NSF Standard 54

- elongation change at yield no more than 20% in either direction, and elongation value equal to or greater than the minimum as-received value in the material property table of NSF Standard 54

- tear resistance at least 80% of the initial value, and equal to or greater than the minimum as-received value in the material property table of NSF Standard 54

- modulus of elasticity at least 70% of the initial value, and equal to or greater than the minimum as-received value in the material property table of NSF Standard 54.

In general, the specified limits on property changes were used rather than the comparison to minimum properties. This was done because the methods used for measuring elongation and elastic modulus differed from those used to specify minimum properties [9], so these values could not be compared. Specific exceptions to the NSF-proposed requirements are discussed in the results section, where appropriate.

For the non-HDPE liners, stability of properties was the primary acceptance criteria. Also, a ±25% criteria was used for reporting purposes based on the mean acceptance criteria for all flexible membrane liners tested by NSF [8].

Acceptance criteria for radiation compatibility have not been developed. However, since the lifetime dose can be simulated by accelerating the exposure rate, stability of properties should not be considered as a criterion. Instead, the criteria should be based on the required properties at the end of the service period. A shielding calculation based on the radioactive inventory in the waste has shown that the 30-year dose to the liner in the current DSSF disposal scenario is only 14 rad [3]. Although direct exposure of the liner to any leachate would cause some increase in the dose, it is expected that the increase in dose would be less than two orders of magnitude. As a result, the effect of radiation exposure on the liner properties is expected to be minimal.

TEST RESULTS

Failed Liners

    After 60 days of immersion, the EIA liner was judged to be incompatible due to swelling, weight gain, and 40% reduction in the measured break strength and elongation. Results from the radiation tests showed that the EIA liner is reasonably resistant to radiation damage, with puncture elongation exhibiting the greatest change in value (a 36% decrease) between the 0.6 Mrad and the 3.6 Mrad dose. The change was then stable at doses of 16.1 and 38.9 Mrad.

    The polyurethane liner lost approximately 90% of its break strength after only 30 days, after which testing was stopped. It is believed that the high temperature and pH of the waste caused hydrolysis of esters in the polyurethane structure.

    A nonreinforced Hytrel polyester was immersed in DSSF at 90°C. Within 13 days the liner had completely dissolved, probably because of hydrolysis of esters caused by the high temperature and high pH of the waste.

HDPE and Polypropylene Chemical Compatibility

    Chemical compatibility results for HDPE and polypropylene samples immersed at 90°C are presented in Tables 2 and 3, respectively. Ambient temperature data are omitted since the effect of immersion at ambient temperature is generally less severe than at 90°C. Total organic carbon (TOC) of the immersion solution was measured every 30 days to determine if replacement of the waste was required. Since decreases in the waste TOC level were not observed, the same waste was used for the entire test.

    During the compatibility tests, the temperature of each immersion cell was measured hourly to determine if the cell temperatures were being controlled within a ±2°C temperature range, as called for in EPA Method 9090 [2]. The temperature control was outside this specification several times. The loss of temperature control is not thought to have had a significant effect on the results. The time that the cells were below the specified temperature was a small percentage of the total immersion time. Time periods during which the temperature was greater than 92°C should provide conservative results if the liner is found to be satisfactory.

HDPE results: The results from chemical compatibility testing
of the HDPE liner are listed in Table 2. The length, thickness, and
weight of the liner samples were not significantly changed during the
120 days of immersion. The stability of these properties indicates
that the HDPE liner will not absorb components of the waste, which
would result in swelling and an increase in weight. In addition, it
appears that the liner does not shrink or release significant quantities of material. These observations are substantiated by the small
changes observed in the HDPE liner volatile and extractables contents, and the small observed change in liner specific gravity. In
addition, visual observation of the immersed samples and the
environmental stress cracking samples did not indicate any change in
the material after 120 days (2880 hours) of immersion. The results
imply that the liner is resistant to chemical attack by the DSSF at
90°C, and should retain its ability to resist permeation.

The effects of immersion on liner tear strength, elastic modulus,
and hardness do not create a concern. In all cases the properties
stabilized within the NSF-proposed criteria for HDPE liner during the
120 days of immersion testing.

Changes in the liner properties that may be reason for concern
were the liner break strength and elongation and the puncture force
and elongation. These properties do not appear to have stabilized
within the 120-day test period, and therefore are not in accord with
the NSF-proposed criteria for property stability. In all cases the
changes result in improved liner properties. The liner requires more
force to break or puncture and will stretch further before failure.
It is believed that the changes are due to an increase in
crystallinity in the HDPE liner. Previous studies have shown an
increase in the crystallinity of HDPE at elevated temperatures [10].
Therefore, it is believed that the trend toward higher tensile and
puncture forces and greater elongations does not indicate chemical
incompatibility. The HDPE liner appears compatible with the DSSF
slurry at 90°C. The results correspond with other sources that show
HDPE's resistance to chemical attack and degradation at temperatures
under 90°C [11,12,13,14].

Polypropylene results: The results from chemical compatibility
testing of the polypropylene liner are listed in Table 3. The
length, thickness, and weight of the liner samples were not significantly changed during 120 days of immersion. The stability of these
properties indicates that the polypropylene liner will not absorb
volatile or extractable materials that cause the polymeric material
to swell and increase in weight. In addition, it appears that the
liner does not shrink or release significant quantities of material.
These observations are substantiated by the small changes in volatile
and extractables contents, and the small change in specific gravity
during immersion. Visual observations of the polypropylene liner
samples (including the environmental stress cracking samples) found
no sign of chemical degradation in the samples during 120 days
(2880 hours) of immersion. The results imply that the polypropylene
liner is resistant to chemical attack by the DSSF at 90°C, and should
retain its ability to resist permeation.

TABLE 2 -- 120-day test results for HDPE liner immersed in DSSF at 90°C

Property and Units		Non-Irradiated Liner Samples			
		30 Day	60 Day	90 Day	120 Day
Tear Strength (% change)[a]	-md-[b]	-3.0	0.9	0.0	1.9
	-td-[b]	-4.8	-4.5	6.2	-0.4
Specific Gravity (% change)		0.8	0.3	0.3	0.6
Puncture Force (% change)		-3.1	1.5	3.6	9.1
Puncture Elongation (% change)		-3.7	14.9	22.2	33.5
Hardness, Duro D units (unexposed liner 60.2)		61.0	57.9	61.7	59.5
Dimensional Change (%)[c] (l,w,thk,wt)		≤0.5	≤0.5	≤0.8	≤0.8
Volatiles (%) (unexposed liner = 0.250%)		0.199	0.170	0.200	0.177
Extractables (%) (unexposed = 0.132%)		0.362	0.412	0.804	0.673
Environmental Stress Cracking		-no-[d]	-no-	-no-	-no-
Tensile Properties:[e]					
Yield Strength (% change)	-md-	6.0	1.1	7.6	5.4
	-td-	2.9	0.5	2.0	2.5
Yield Elongation (% change)	-md-	12.1	10.6	12.1	15.2
	-td-	4.5	0.0	6.1	4.5
Break Strength (% change)	-md-	-1.6	-0.1	-3.2	10.5
	-td-	7.6	7.6	2.8	25.2
Break Elongation (% change)	-md-	2.7	3.0	5.2	12.7
	-td-	9.2	8.2	1.2	21.5
Stress at 100% Elongation (% change)	-md-	10.1	9.5	10.2	9
	-td-	2.1	0.7	3.5	1.7
Stress at 200% Elongation (% change)	-md-	4.2	0.5	5.8	2.7
	-td-	3.4	1.0	5.2	3.7
Elastic Modulus (% change)	-md-	-18.5	-1.3	-1.1	-5.4
	-td-	-9.6	-0.3	-15.9	-0.7

(a) Compared to the measured value for unexposed specimens.
(b) -md- machine direction, -td- transverse direction.
(c) Values indicate the magnitude of the largest change.
(d) -no-: no stress cracking observed.
(e) Tensile values calculated from 5 unexposed and 3 exposed specimens at each immersion period. For individual measurement values and number of specimens for all tests, see Farnsworth and Hymas 1989 [4]. Elongations determined from grip separation.

TABLE 3 -- 120-day test results for polypropylene liners immersed in DSSF at 90°C

Property and Units		Non-Irradiated Liner Samples			
		30 Day	60 Day	90 Day	120 Day
Tear Strength (% change)[a]	-md-[b]	-0.6	9.1	8.4	4.4
	-td-[b]	3.6	14.0	11.9	7.3
Specific Gravity (% change)		0.1	0.4	0.1	-0.1
Puncture Force (% change)		-2.5	-0.4	-2.7	-11.6
Puncture Elongation (% change)		-40.1	-38.7	-31.8	-17.4
Hardness, Duro D units (unexposed 68.6)		66.8	69.3	69.1	67.3
Dimensional Change (%)[c] (l,w,thk,wt)		<0.6	<0.6	<0.6	<1.1
Volatiles, wt% of Sample (0.01% unexposed)		0.0779	0.628	0.1112	0.0132
Extractables (wt%) (unexposed = 0.405%)		0.397	0.599	0.423	0.727
Environmental Stress Cracking		-no-[d]	-no-	-no-	-no-
Tensile Properties:[e]					
Yield Strength (% change)	-md-	-8.6	-4.0	-8.7	-4.8
	-td-	-11.5	-7.9	-9.9	-4.4
Yield Elongation (% change)	-md-	36.8	35.4	23.7	46.3
	-td-	41.8	37.3	31.7	21.1
Break Strength (% change)	-md-	10.7	13.6	12.2	12.2
	-td-	2.8	2.3	6.1	10.1
Break Elongation (% change)	-md-	104.1	112	129.7	30.0
	-td-	29.8	-6.7	-11.8	14.1
Stress at 100% Elongation (% change)	-md-	21.9	22	15.6	14.4
	-td-	3.3	13.3	6.6	20.6
Stress at 200% Elongation (% change)	-md-	5.1	4.2	2.6	6.6
	-td-	0.9	2.1	5.5	8.6
Elastic Modulus (% change)	-md-	-32.9	-29.1	-26.2	35
	-td-	-37.6	-32.8	-31.5	21

(a) Compared to the measured value for unexposed specimens.
(b) -md- machine direction, -td- transverse direction.
(c) Values indicate the magnitude of the largest change.
(d) -no-: no stress cracking observed.
(e) Tensile values calculated from 5 unexposed and 3 exposed specimens at each immersion period. For individual measurement values and number of specimens for all tests, see Farnsworth and Hymas 1989 [4]. Elongations determined from grip separation.

However, five properties (yield elongation, break elongation, elastic modulus, puncture force, and puncture elongation) were outside of the qualitative criteria that were developed for nonirradiated, non-HDPE liner material. One problem encountered was that the thickness of the polypropylene was not uniform. The actual thickness of the nominally 4.3-mm liner varied from 4.0 mm to 5.6 mm. The dimensional samples for ambient and 90°C had thicknesses of 4.7 ± 0.7 mm and 4.8 ± 0.4 mm, where ± represents one standard deviation. The nonuniformity in liner thickness contributed to variability in the results of mechanical property tests.

The yield elongation of the polypropylene liner increased by 40% after 30 days of immersion but remained stable for the remaining 90 days of immersion. It is believed that the increase is a direct result of the temperature of the DSSF slurry, not an indication of degradation. Since the yield elongation stabilized in a beneficial direction, the concern over this deviation was minor. The liner elastic modulus decreased 30% to 35% during the first 30 days of immersion testing, but remained stable thereafter. This decrease is caused by the increase in yield elongation of the liner over the same 30-day period. Likewise, concern over this deviation is minor.

The break elongations of the polypropylene liner varied from +130% to -12% during the 120 days of 90°C immersion. The variations did not appear to be due to liner thickness, but rather a poor repeatability of the measurement. As an example, for unexposed samples, the percent elongation-at-break measurements were 344, 187, 491, 266, 251 in the machine direction and 790, 850, 191, 582, 671 in the transverse direction. Since data from immersed samples was taken in triplicate, changes in properties were masked by the variability in the measurement. From visual observations, it appeared that the liner would thin and break unpredictably during tensile testing.

The liner puncture force appeared to decrease 9% without stabilizing over the last 30 days of immersion. However, this change was clearly correlated to the measured thickness of the specimens before testing. The puncture elongation showed a significant fluctuation after 30 days and then drifted back towards the unexposed value. The puncture elongation was more repeatable than break elongation and the variations were not correlated to the thickness of individual specimens. Although puncture elongation causes some concern, it is suspected that the change in puncture elongation is also related to the unpredictable elongation behavior of the polypropylene liner, and it was concluded that the changes did not indicate incompatibility. All specimens for one immersion period were cut from the same coupon, and there may be variability between coupons.

Although the HDPE liner appears to be superior from the standpoint of chemical compatibility, it is concluded that polypropylene liner may also be suitable for the application. The property changes of the polypropylene liner, which appeared more significant than those for the HDPE liner, were masked by variation in thickness and a lack of repeatability of some measurements. It is suggested that testing be performed using more uniform polypropylene liner material.

Others have shown polypropylene to be more sensitive to thermal oxidative breakdown than polyethylene because of the carbon-hydrogen (C-H) linkages with tertiary carbon atoms [15].

Radiation Effects

High-density polyethylene: The results from radiation compatibility tests on HDPE samples are listed in Table 4. The table headings indicate the nominal dose to the liner, followed by the total immersion time for the sample in DSSF at 90°C. Radiation tests were not performed for ambient immersions. The effects of radiation on the liner can be seen by comparing the 30-day immersion data at different doses for each property. The effect of radiation on sample mass, dimensions, and volatile content is very small. The yield strength increases and yield elongation decreases with increasing dose. The most severe change in properties is a reduction in break strength and elongation of the samples. Also, significant changes were observed in the stress at 100% and 200% elongation at the 38.9 Mrad dose. These changes are believed to be due to crosslinking in the irradiated HDPE [16]. The puncture force increased, as would be expected. However, there was an increase in puncture elongation above what occurs in the absence of radiation. The reason for this is unknown, but does not cause concern since these changes represent 30-year-life changes. Environmental stress cracking was not performed as part of this testing, but previous work with HDPE using a more concentrated DSSF at 75°C showed no stress cracking when the samples were subjected to 0.93 Mrad over a 120-day (2880 hour) period. The effect of radiation on stress cracking has also been examined by others [17].

The 120-day immersion data are for samples irradiated at 30 days and then reimmersed and tested at 120 days of total immersion. In general, the 38.9 Mrad/120-day data are similar to the 38.9 Mrad/30-day data, indicating that radiation does not make the liner more susceptible to chemical degradation. Therefore, radiation and chemical degradation can be evaluated separately. It is concluded that the HDPE liner is resistant to radiation damage and is suitable for the environment of a grout vault.

Polypropylene results: The results from radiation compatibility tests of polypropylene liner are listed in Table 5. As with HDPE, radiation exposure did not change the sample dimensions. Significant changes were observed in the liner's tensile properties, tear strength, and puncture elongation at exposures of 16.1 Mrad and 38.9 Mrad. In general, the effect of radiation on a liner's mechanical properties appears to be greater for polypropylene than for HDPE. This observation is in accordance with previous studies that have shown polypropylene to be less resistant to radiation damage than HDPE. The decrease may be due to the increased amount of chain fracturing that occurs as the polypropylene cross-links during irradiation. In contrast, irradiated HDPE cross-links with very little main chain fracture [18].

TABLE 4 -- Results for irradiated, 120-day HDPE liner in DSSF at 90°C

Property and Units		Radiation Dose (Mrad)/Immersion Time(days)							
		0/ 30	0.6/ 30	3.6/ 30	16.1/ 30	38.9/ 30	0/ 120	0.6/ 120	38.9/ 120
Tear Strength -md-[b]		-3.0	-3.8	-6.2	-5.9	-10.7	1.9	0.3	-13.5
(% change)[a] -td-[b]		-4.8	3.7	6.8	4.4	4.9	-0.4	-1.7	-6.7
Puncture Force (% change)		-3.1	5.0	3.9	7.0	13.0	9.1	8.4	16.2
Puncture Elong. (% change)		-3.7	7.1	66.7	14.9	24.2	33.5	39.8	69.3
Hardness, Duro D units (unexposed 60.2)		61.0	58.2	56.9	59.1	61.4	59.5	60.9	63.7
Dimensional (% change)[c] (l,w,thk,wt)		≤0.5	≤0.7	≤1.2	≤0.3	≤0.9	≤0.8	≤0.7	≤0.5
% Volatiles (wt%) (unexposed = 0.250%)		0.20	0.34	0.22	0.19	0.18	0.18	0.17	0.16
Tensile Properties:[d]									
Yield Strength -md-		6.0	3.4	7.4	12.7	16.2	5.4	6.7	7.0
(% change) -td-		2.9	1.8	6.7	11.1	11.9	2.5	0.5	6.2
Yield Elongation -md-		12.1	18.2	10.6	4.5	3.0	15.2	10.6	3.0
(% change) -td-		4.5	9.1	2.3	-3.0	7.6	4.5	4.5	-1.5
Break Strength -md-		-1.6	-2.2	-3.4	-14.4	-29.4	10.5	-5.6	-26.1
(% change) -td-		7.6	-1.7	3.8	-30.4	-21.4	25.2	9.9	-24.1
Break Elongation -md-		2.7	3.7	-0.9	-30.4	-69.6	12.7	-6.1	-60.8
(% change) -td-		9.2	4.1	6.8	-48.1	-56.6	21.5	7.7	-65.9
Stress at 100% -md-		10.1	7.5	9.2	11.8	27.0	9.0	5.4	27.4
Elongation (% -td- change)		2.1	3.5	7.5	17.9	27.8	1.7	2.8	23.7
Stress at 200% -md-		4.2	1.3	3.6	13.1	23.6	2.7	2.1	20.1
Elongation (% -td- change)		3.4	3.3	7.1	19.0	28.0	3.7	3.7	26.9
Elastic Modulus -md-		-18.5	-11.8	1.6	5.9	1.4	-5.4	-0.2	1.0
(% change) -td-		-9.6	-10.7	-1.3	21.2	0	-0.7	-9.0	14.3

(a) Compared to the measured value for unexposed specimens.
(b) -md- machine direction, -td- transverse direction.
(c) Values indicate the magnitude of the largest change.
(d) Tensile values calculated from 5 unexposed and 3 exposed specimens at each immersion period. For
     individual measurement values and number of specimens for all tests, see Farnsworth and Hymas 1989 [4].
     Environmental stress cracking and extractables not performed on irradiated samples. Related work
     mentioned in text. Elongations determined from grip separation.

TABLE 5 -- Results for irradiated, 120-day polypropylene liner in
DSSF at 90°C

Property and Units		Radiation Dose (Mrad)/Immersion Time(days)							
		0/ 30	0.6/ 30	3.6/ 30	16.1/ 30	38.9/ 30	0/ 120	0.6/ 120	38.9/ 120
Tear Strength -md-[b]		-0.6	2.1	-1.5	-11.8	-31.1	4.4	6.9	-27.6
(% change)[a] -td-[b]		-3.6	4.2	-2.0	-12.5	-28.8	7.4	4.3	-29.9
Puncture Force (% change)		-2.5	5.0	0.8	2.4	16.4	-11.6	22.4	12.7
Puncture Elong (% change)		-40.1	-41.2	-44.3	-11.7	-44.9	-17.4	-23.6	-16.1
Hardness, Duro D units (unexposed 60.2)		66.8	70.3	67.1	67.7	70.6	67.3	68.7	72.5
Dimensional (% change)[c] (l,w,thk,wt)		≤0.6	≤0.3	≤0.4	≤0.4	≤0.2	≤1.2	≤0.3	≤0.2
% Volatiles (wt%) (unexposed = 0.01%)		0.08	0.05	0.05	0.09	0.11	0.01	0.05	0.08
Tensile Properties:[d]									
Yield Strength	-md-	-8.6	1.8	-14.8	-3.6	8.6	-4.8	-3.6	-7.7
(% change)	-td-	-11.5	-2.0	-6.5	-5.8	6.6	-4.4	-2.2	1.1
Yield Elongation	-md-	36.8	11.4	14.3	43.6	23.7	46.3	33.9	-14.8
(% change)	-td-	41.8	9.2	11.3	11.3	11.3	21.1	28.2	63.9
Break Strength	-md-	10.7	5.8	-14.0	-50.5	6.3	12.2	10.3	-16.4
(% change)	-td-	2.8	2.3	-21.8	-4.8	-4.9	10.1	2.0	29.9
Break Elongation	-md-	104.1	-44.1	58.9	-60.3	-73.6	30.0	48.5	-95.3
(% change)	-td-	29.8	-34.1	-12.9	-71.1	-86.3	14.1	-70.8	-89.2
Stress at 100%	-md-	21.9	4.4	-0.5	-26.0	(e)	14.4	19.4	(e)
Elong. (% change)	-td-	3.3	11.6	1.9	-4.3	(e)	20.6	4.1	(e)
Stress at 200%	-md-	5.1	2.5	-13.8	(e)	(e)	6.6	6.3	(e)
Elong. (% change)	-td-	0.9	6.1	-6.5	(e)	(e)	8.6	(e)	(e)
Elastic Modulus	-md-	-32.9	-8.6	-25.4	-27.5	-12.3	-35.0	-28.1	--
(% change)	-td-	-37.6	-10.0	-16.1	-15.4	-4.3	-21.0	-23.7	-38.3

(a) Compared to the measured value for unexposed specimens.
(b) -md- machine direction, -td- transverse direction.
(c) Values indicate the magnitude of the largest change.
(d) Tensile values calculated from 5 unexposed and 3 exposed specimens at each immersion period. For
     individual measurement values and number of specimens for all tests, see Farnsworth and Hymas 1989 [4].
     Environmental stress cracking and extractables not performed on irradiated samples. Elongations
     determined from grip separation.
(e) Samples broke before reaching this elongation value.

The break elongation was the most severely affected property. The average break elongation after 120 days immersion and a 38.9 Mrad dose was reduced to only 8% of the elongation of the unexposed liner. Therefore, a 30-year life for this liner in the interior environment of the grout vault is not supported by the data. It is questionable if a 30-year life is required in the disposal scenario since the waste solidifies quickly and leachate is not expected. However, due to design changes, the actual expected 30-year dose is only 14 rad, which should not have any appreciable effect on the liner.

Data from samples irradiated after 30 days of immersion and then reimmersed for an additional 90 days were examined to determine if radiation affected the chemical resistance of the liner. There appears to be some effect on the yield elongation and elastic modulus values.[a] However, in general, it appears that radiation does not significantly affect the chemical degradation rate. Therefore, the chemical and radiation results can be considered separately in this case.

REFERENCES

[1] Resource Conservation and Recovery Act of 1976. Public Law 94-580, 42 USC 6901-6907 et. seq., 1976.
[2] Environmental Protection Agency. 1986. "Method 9090: Compatibility Test for Wastes and Membrane Liners." In Test Methods for Evaluating Solid Waste, Vol. 1A: Laboratory Manual, Physical/Chemical Methods. 3rd ed., SW-846. Environmental Protection Agency, Washington, D.C.
[3] Richmond, W. G. 1988. Methods and Data for Use In Determining Source Terms for the Grout Disposal Program. SD-WM-TI-355, Westinghouse Hanford Company, Richland, Washington.
[4] Farnsworth, R. K., and C. R. Hymas. 1989. The Compatibility of Various Polymeric Liner and Pipe Materials with Simulated Double-Shell Slurry Feed at 90°C. PNL-6969. Pacific Northwest Laboratory, Richland, Washington.

[5] Bolt, R. O. and J. G. Carroll. 1963. "General Radiation Effects of Elastomeric Material." In Radiation Effects on Organic Materials, pp. 252-256. Academic Press, New York, New York.
[6] Harrington, R. 1957. "Elastomers for Use in Radiation Fields (Part 1)," Rubber Age 81(6):971-980.

[7] Mattia, R. F. and D. R. Luh. 1971. "Radiation Resistance of EPDM, Polychloroprene, and Chlorosulfonated Polyethylene." Contribution No. 286, presented at the 1971 Tri-State Regional Meeting of the Massachusetts, Connecticut, and Rhode Island Rubber Groups, May 13, 1971, Auburn, Massachusetts.

(a) Elastic modulus for polypropylene liner was calculated from yield strength and elongation using ASTM D638.

[8] Bellen, G., R. Corry, and M. L. Thomas. 1988. Development of Chemical Compatibility Requirements for Assessing Flexible Membrane Liners, Volume 1. Prepared for the U.S. Environmental Protection Agency by National Sanitation Foundation, Ann Arbor, Michigan.

[9] National Sanitation Foundation. 1985. Standard Number 54: Flexible Membrane Liners. Rev. Standard. National Sanitation Foundation, Ann Arbor, Michigan.

[10] Mitchell, D. H. 1986. Geomembrane Selection Criteria for Uranium Tailings Ponds. NUREG/CR-3974 (PNL-5224), Prepared for the U.S. Nuclear Regulatory Commission by Pacific Northwest Laboratory, Richland, Washington.

[11] Wallder, V. T., W. J. Clarke, J. B. DeCoste, and J. B. Howard. 1950. "Weathering Studies on Polyethylene: Wire and Cable Applications." In Industrial and Engineering Chemistry 42(11): 2320-2325.

[12] Biggs, B. S. 1953. "Aging of Polyethylene." In Proceedings of the NBS Semicentennial Symposium on Polymer Degradation Mechanisms, pp. 137-145. National Bureau of Standards Circular 525, Washington D.C.

[13] Gordon, N. R. 1966. "The Effects of Radiation on Plastics." BNWL-SA-659, Pacific Northwest Laboratory, Richland, Washington.

[14] Sweeting, O. J. 1971. "Polyethylene." In The Science and Technology of Polymer Films, pp. 132-183. Wiley-Interscience, New York, New York.

[15] Cooke, T. F., and L. Rebenfeld. 1988. "Effect of Chemical Composition and Physical Structure of Geotextiles on Their Durability." In Geotextiles and Geomembranes Vol. 7, pp. 7-22. Elsevier Science Publishers Ltd., England.

[16] Charlesby, A. 1960. "Polyethylene." In Atomic Radiation and Polymers, pp. 198-257. Pergamon Press, New York, New York.

[17] Soo, P. Effects of Chemical and Gamma Irradiation Environments on the Mechanical Properties of High-Density Polyethylene (HDPE). BNL-NUREG--40842, Brookhaven National Laboratory, Upton, New York

[18] Charlesby, A. 1960. "Other Crosslinking Polymers." In Atomic Radiation and Polymers, pp. 323. Pergamon Press, New York, New York.

Richard W. Thomas and Mark W. Cadwallader

SOLVENT SWELL OF POLYETHYLENE BY
THERMOMECHANICAL ANALYSIS

REFERENCE: Thomas, R. W. and Cadwallader, M., "Solvent
Swell of Polyethylene by Thermomechanical Analysis,"
Geosynthetic Testing for Waste Containment Applications,
ASTM STP 1081, Robert M. Koerner, Ed., American Society
for Testing and Materials, Philadelphia, 1990.

ABSTRACT: An important property of high density polyethylene
(HDPE) for lining waste management facilities is how it behaves when
exposed to chemicals. A new method to study this has been developed
with the use of a thermomechanical analyzer (TMA). The technique
monitors the size of a sample as a function of time while it is exposed
to a challenge chemical. The resulting absorption curve can be used to
generate the coefficient of diffusivity and the percentage swell at
equilibrium. Four different densities of PE were studied by this
method. The results showed that the diffusivity, solubility and
permeability of a PE resin are dramatically affected by the resin
density.

KEYWORDS: Polyethylene, Diffusion, Permeation, Solvent Swell,
Thermomechanical Analyzer.

INTRODUCTION

Because the primary function of a synthetic liner in waste
containment is to prevent migration of chemicals into the groundwater,
properties of molecular movement through the liner such as chemical
absorption, desorption, swelling and permeation are important. Among
polyethylene liners the level of crystallinity (measured indirectly by the
density) is very important in preventing molecular movement through
the liners. Exactly how much this molecular movement depends on
density is an important question which can be very well addressed with
the techniques of thermomechanical analysis.

Permeation is the process in which permeant molecules migrate
through a membrane. Mathematically, it is the product of two separate
processes, diffusivity and solubility.

Mr. Thomas is Director of Laboratories at TRI Environmental, Inc., 9063 Bee
Caves Road, Austin, Texas 78733; Mr. Cadwallader is Director of Research and
Technical Development at Gundle Lining Systems, Inc., 1340 E. Richey Road, Houston,
Texas 77073.

Diffusion can be thought of as the rate of permeant migration into the membrane in the absence of dimensional changes. Usually, this involves travel through micro-voids or potential voids. Potential voids are those which are continuously created and destroyed as the penetrant pushes through the membrane. Conversely, solubility is the process of creating new spaces for the penetrant to occupy by swelling of the membrane.

In terms of mass uptake, the classic description of diffusivity is given by Fick's equation [1]:

$$D = \left(\frac{h\theta}{4M} \right)^2$$

where

D = diffusivity
h = initial thickness
M = mass uptake at equilibrium, and
O = slope of the plot of % weight gain vs. square root time

This equation has been used sucessfully to measure diffusivity and solubility by mass uptake [1-3].

In order to evaluate diffusion in terms of solvent swell, one must consider the coefficient of hygroelasticity, [3].

where
$$\mu = \frac{\Delta L / L_o}{\Delta W / W_o}$$

L = thickness and
W = weight

However, if one assumes that the ratio of thickness change to weight change stays constant throughout the swell experiment, the thickness can be substituted for the weight in Fick's equation.

The use of a thermomechanical analyzer (TMA) for measuring the swell and dissolution of thin polymer films was described in 1970 [4]. The results of that study showed that for the systems evaluated, Fickian diffusion occurred up to about 60% of the equilibrium swell volume. Another study [3] showed a direct correlation between mass gain and swell by TMA, for polyurethanes. Therefore, the model based on Fick's equation was chosen to evaluate the effect of chlorobenzene on a series of polyethylenes which differed in density.

EXPERIMENTAL

Resins

The polyethylene resins used in this study were neat (no fillers) and had different densities. The resins were molded into small discs 6.3mm in diameter and 1mm in thickness. They were first thermally annealed in a platen press and then chemically annealed in chlorobenzene. The densities of the annealed discs were determined with the use of a gradient density column.

Swell Experiment

A DuPont model 943 TMA was used with the following modifications. First, a glass test tube was constructed to fit between the quartz sample holder tube and the furnace. Second, a length of small diameter Teflon tubing was inserted into the sample holder tube from the back alongside the thermocouple. The tubing was connected to a syringe from which the challenge chemical could be transferred into the sample compartment.

The experiment was started by first loading the sample disc into the sample holder, lowering a 6.3mm probe onto the sample and loading the probe with a 10g weight. Next, the test tube and furnace were positioned around the sample holder tube and the sample was then heated to 29°C. Once the temperature had equilibrated, chlorobenzene, which was preheated to about 30°C, was introduced to the sample compartment through the Teflon tubing with a syringe. The instrument was then quickly zeroed and the experiment started. The data collected was the linear displacement and the temperature as a function of time.

Each of the four PE resins were tested in triplicate.

RESULTS

The discs that were evaluated had densities of 0.951, 0.940, 0.925 and 0.903 g/cm^3 as determined in a density gradient column. The resins were treated thermally and chemically to ensure that no thermal stresses were present and that each resin was at its natural state of crystallinity.

The resin pellets were molded into discs in a platen press at 158°C. After the pellets were placed into the mold, the mold was closed and heated at temperature for five minutes. A pressure of 9072 Kg was then applied and the platens held at temperature for five minutes more. The press was then turned off and the samples were allowed to cool under pressure to room temperature overnight. During the initial stage of this study, it was discovered that the nine chambers in the mold did not cool at the same rate. This showed up as a reproducible dip in the % swell vs. square root time plots.

runs of the same samples. Because of these results, each disc was soaked in chlorobenzene for 24 hours and dried in a vacuum oven at 30°C for 24 hours. This also demonstrates the utility of this technique for detecting thermal stresses in polyethylene.

A typical absorption curve for a 0.951 g/cm³ resin is shown in Figure 1. Notice, that the temperature quickly reaches 29°C and holds within 0.3°C. The small pertubations in the swell curve were removed by a smoothing routine which time averaged the data.

Figure 2 is a summary plot of representatives of each of the four densities. As seen in the plot, the percentage swell at equilibrium was greatly increased as the density decreased.

After the raw data was obtained, plots of % swell vs. square root time were prepared and the slopes and diffusivities were calculated. Table 1 lists the results of triplicate experiments for each density of resin.

Table 1 -- Swell and Diffusion Parameters

Density, g/cm³	Slope[a], %/sec $^{1/2}$	Equil. swell, %	Diffusivity, cm²/sec x 10⁷
0.951	0.022	1.97	2.35
0.940	0.027	2.11	3.13
0.925	0.066	4.67	3.90
0.903	0.271	13.97	7.22

[a] slope from 0 to 60% of equilibrium

Notice that the % swell seems to be more sensitive to density than the diffusivity. The diffusivity of 2.35×10^{-7} cm²/sec for the 0.951 g/cm³ resin agrees well with 1.0×10^{-7} cm²/sec for HDPE and methylene chloride reported previously [2]. Additionally, the relationship between diffusion and density is linear in a log-log plot as reported elsewhere [5].

Since permeation is the product of diffusivity and solubility, the relative effect of density on permeability can be implied by these results. This is shown in Table 2.

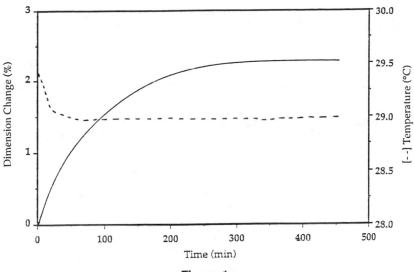

Figure 1
Absorption curve of a typical 0.951 g/cm³ HDPE sample

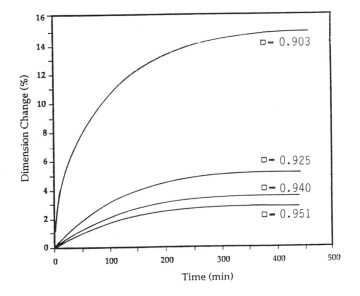

Figure 2
Absorption curves of four densities of HDPE

Table 2 -- The Relative Change in
Difficulty, Solubility and Permeability

Density, g/cm^3	Difficulty	Solubility	Permeability
0.951	1.0	1.0	1.0
0.940	1.3	1.1	1.4
0.925	1.7	2.4	4.1
0.903	3.1	7.1	22.0

As seen in the table, a change in density from 0.951 g/cm^3 to 0.903 g/cm^3 produced an increase in permeability of 22 times. These results are similar to others which measured oxygen permeability as a function of density [6].

CONCLUSIONS

The results of this work showed that a modified TMA can be used to study the interaction between chemical agents and polyethylene. The method can also be used to detect small thermal stresses and can generate coefficients of diffusivity which generally agree with published results.

Finally, the relationship between polyethylene density, and swell and diffusion has been established. These results are important for design considerations because permeation is strongly dependent on density in polyethylene. High density polyethylene does a better job of preventing permeation. Barrier layers of high density polyethylene should therefore be incorporated where maximum barrier action is desired.

REFERENCES

[1] Aminabhavi, T.M., Thomas, R.W. and Cassidy, P.E., "Predicting Water Diffusivity in Elastomers", Polymer Engineering and Science, Vol. 24, No. 18, Dec. 1984, pp. 1417-1420.

REFERENCES, cont'd

[2] Britton, L.N., Ashman, R.B., Aminabhavi, T.M., and Cassidy, P.E., "Prediction of Transport Properties of Permeants through Polymer Films: A Simple Gravimetric Experiment", Journal of Chemical Education, Vol. 65, No. 4, 1988, pp. 368-370.

[3] Aithel, U.S., Aminabhavi, T.M. and Thomas, R.W., "Interaction of Organic Solvents with Polyurethane Membranes", Rubber Chemistry and Technology, 1989, in press.

[4] Machin, D. and Rodgers, C.E., "Isothermal Kinetic Studies Using a Thermomechanical Analyzer: 1. Rates of Polymer Swelling and Dissolution", Polymer Engineering and Science, Vol. 10, No. 5, 1970, pp. 300-304.

[5] Fleischer, G., "A Pulsed Field Gradient NMR Study of Diffusion in Semi-Crystalline Polymers: Self-Diffusion of Alkanes in Polyethylenes", Colloid & Polymer Science, Vol. 262, 1984, pp. 919-928.

[6] Geetha, R., Torikai, A., Yoshida, S., Nagaya, S., Shirakawa, H., and Fueki, K., "Radiation-Induced Degradation of Polyethylene: Effect of Processing and Density on the Chemical Changes and Mechanical Properties", Polymer Degradation and Stability, Vol. 23, 1988, pp. 91-98.

Ian D. Peggs and Daniel S. Carlson

THE EFFECTS OF SEAMING ON THE DURABILITY
OF ADJACENT POLYETHYLENE GEOMEMBRANES

REFERENCE: Peggs, I.D. and Carlson, D.S., "The Effects of
Seaming on the Durability of Adjacent Polyethylene
Geomembranes," Geosynthetic Testing for Waste Containment
Applications, ASTM STP 1081, Robert M. Koerner, Ed., American
Society for Testing and Materials, Philadelphia, 1990.

ABSTRACT: Several stress cracking failures in the liners of
fluid waste impoundments have been observed in and adjacent to
all types of seams in polyethylene geomembranes. The majority
of these cracks occur at the outer edge of extruded fillet and
lap seams and the inside edge of hot wedge seams. Stress
cracking is the predominant mechanism that presently
compromises the long-term durability of polyethylene
geomembrane and that, in cold weather, may initiate
spectacular, rapid crack propagation failures. Each type of
polyethylene resin has its own fundamental stress cracking
characteristics, which can be modified by the time/temperature
history during manufacture of the geomembrane and,
subsequently, during seaming. This paper reports the results
of a preliminary series of constant tensile load tests using
single edge notched specimens to determine the effects on
stress crack growth rates of different types of seams in
different polyethylene geomembranes. The influence of
microstructure on stress cracking is also assessed.

KEYWORDS: stress cracking, polyethylene, geomembrane, seams,
liners, crack growth rates, brittle failures

INTRODUCTION

Stress cracking (SC) has been observed in polyethylene (PE) natural
gas distribution pipe [1] and in PE geomembranes [2,3]. This is
perhaps not surprising since many geomembrane and pipe resins are from

Dr. Peggs is president and Mr. Carlson is physical laboratory
technician of GeoSyntec, Inc., 3050 S.W. 14th Place, Suite 18, Boynton
Beach, FL 33426.

the same families or they may even be identical. Only recently have resins been specifically developed for geomembrane applications.

Presently, SC appears to be the factor that predominantly compromises the durability of PE geomembranes. It has been observed most often on the side slopes of uncovered liquid impoundments, in most manufacturers' materials, and in all types of seams and occurs predominantly at the following locations:

- Extruded fillet seams
  -- in the bottom sheet at the edge of the seam
  -- through the center of the seam

- Extruded lap seams
  -- in the bottom sheet at the edge of the seam
  -- through the center of the seam
  -- in the top sheet at the edge of the seam

- Double hot wedge seams
  -- in the top sheet at the outer edge of the seam
  -- through the center of one track of the seam
  -- in the bottom sheet at the outer edge of the seam

Many stress cracks either have occurred or have initiated at places where excessive amounts of thermal energy have been put into the seam/geomembrane area. The most predominant of these areas is where an initial seam containing a defect has been repaired by placing an additional extruded bead on top of the initial seam or along the edge of it. Similarly, cracks have initiated at seam intersections and where large volumes of extrudate have been deposited in error.

In all such instances, of course, a stress was needed to initiate the cracking, and generally the stress was provided by the thermal contraction of the exposed geomembrane at low winter and spring temperatures.

Furthermore, early in 1989, the existence of a new phenomenon in PE geomembranes was confirmed, one that had not been seen in the pipe industry but one that has been expected and researched: rapid crack propagation (RCP). In three western states and one Canadian province, exposed liners shattered like glass over significantly large areas. Such RCP cracks in pipe had velocities of over 350 m/s [4].   After investigating these failures, it is the authors' opinion that these spectacular, RCP failures occur only if a pre-existing stress crack of critical geometry is caused to grow at a critical rate.   Also, if a stress crack is not present, RCP will likely not occur under the influence of temperature and stress alone. Therefore, it is essential that we understand how seaming affects the microstructure, SC resistance, and therefore durability of adjacent geomembranes.

MULTIAXIAL STRESS BREAKS

First, it should be stated that the brittle cracking (SC) that occurs at the edges of seams is not the type of fracture that occurs when a seam is simply stressed beyond its yield point under the biaxial or plane strain conditions that occur in the field. Experimental work has shown that the tensile testing of wide strip seam specimens and the burst testing of large diameter, 60 cm, seam specimens still produce ductile breaks in the geomembrane, or adjacent to the seam, even though the overall stress/strain curve does not show a distinct yield point [5]. The distinctly brittle cracks seen in the field are clearly a different phenomenon.

SEAM FABRICATION AND STRUCTURE

The production of a seam involves the input of thermal energy to two interfaces, the mixing of molten material, and its solidification to produce a material with the same basic microstructure as the parent material. However, the mixed zone will not have the same microstructural orientation as the original geomembrane, nor will it have the same oxidation resistance since some antioxidant would have been consumed in the seaming process.

A "pool" of structurally modified material, such as shown in Figure 1, can be envisaged to occur in the geomembrane immediately underneath the seam. In addition, Figure 1 shows that there will be a heat-affected zone adjacent to this "pool" where the geomembrane has been heated significantly but does not become liquid. In this area, the additional thermal energy (annealing) may cause secondary crystallization and therefore an increase in overall crystallinity. There is little doubt that these effects will interact synergistically to affect the geomembrane microstructure and residual stress distribution underneath and at the edge of the newly formed seam. The field performance of geomembranes supports this premise, as does the work of Charron [6].

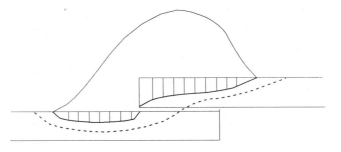

▯▯▯ molten "pool"

‐ ‐ ‐ ‐ boundary of heat-affected zone

FIG. 1 -- Schematic of geomembrane seam.

STRESS CRACK TESTING

The SC resistance of a geomembrane is now commonly assessed by the constant tensile load (CTL) test using single edge notched specimens. The test procedure is outlined in the Geosynthetic Research Institute standard GM-5 [7]. Specimens can be tested to failure to produce the curve shown in Figure 2, or crack growth rates can be measured by the periodic removal of specimens, preparation of microsections [8], and measurement of the extension of the artificial notch. The notch is cut into the specimen to produce the plane strain conditions that exist in the field.

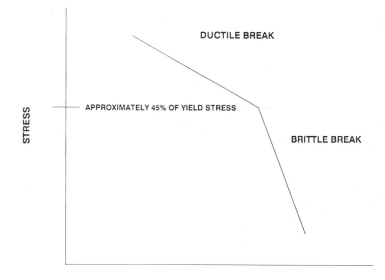

FIG. 2 -- Stress rupture curve from CTL test.

TEST METHODOLOGY AND RESULTS

Figure 3 shows the crack growth rates and failure times measured on all the commonly used PE geomembranes, and some experimental resin geomembranes, tested at a stress of 8.6 MPa (1250 psi) in a 10% Igepal solution at 50°C. There is a factor of 100 difference between the least durable and the most durable commonly used PE geomembranes. The geomembrane manufacturers and resins have not been identified since these preliminary data have been determined on two thicknesses of geomembrane and two of the curves were developed from cast plaques and not extruded geomembrane. Nevertheless, the two extreme curves were measured on geomembrane of nominally the same thickness.

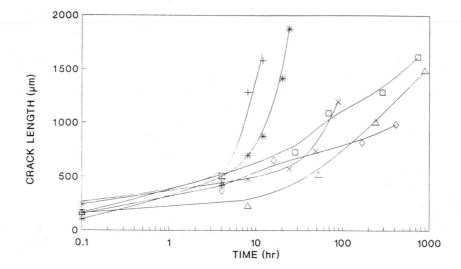

FIG. 3 -- Stress crack growth characteristics of commonly used PE
          geomembranes.

A number of CTL tests were performed with the edge of the seam on the bottom sheet positioned at the center of the ASTM D638, Type IV dogbone specimen. The artificial notch was cut into the top surface of the bottom geomembrane right at the edge of the seam in the same location that field stress cracking occurs.

In performing this seam test, two different approaches to gripping and tensioning the specimen may be taken as shown in Figure 4.   The field conditions may be reproduced if the specimen is gripped on each sheet as shown by A-A.   Alternatively, the specimen is gripped on the bottom sheet on one side of the notch and on the free flap of the bottom sheet on the other side of the notch (B-B).   In the latter way, the additional tensile stresses induced by rotation of the seam in the field-type method do not affect the crack growth rates.   This is a critical factor and one that must be identified when seam performance is compared to geomembrane performance.   The latter method was used in this study, which was intended as a fundamental geomembrane study.

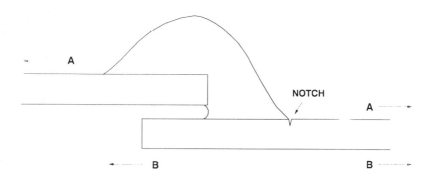

FIG. 4 --   Schematic of CTL test specimen whereby tensioning from A
             to A reproduces field conditions while tensioning from B
             to B provides direct comparison with parent geomembrane
             test.

Table 1 shows a number of test results generated on geomembranes and seams.

TABLE 1 -- Comparison of CTL test crack growth rates and break times for different PE geomembranes and seams

Specimen		Geomembrane		Seam	
Resin	Description	Break Time (hr)	Crack Length[a] (μm)	Break Time (hr)	Crack Length[a] (μm)
A	low carbon	--	17	--	86
A	low carbon, aged	--	34	--	--
B	typical	--	26	--	36, 226[b]
A	old resin	--	120	--	143[c]
A	new resin	--	38	--	203
A	intermediate resin	--	128	--	103
C	improved "original" resin	--	41	--	149, 205
A	aged	202	--	198	--
A	aged	--	383	--	1208
D	fillet seam	200	--	120	--
D	hot wedge seam	200	--	145	--
D	textured, fillet seam	300	--	130	--
D	textured, wedge seam	300	--	160	--

[a] After the same time in both geomembrane and seam specimens.
[b] Contained residual stress.
[c] Seamed geomembrane was a different thickness to the nonseamed geomembrane.

It appears that in the majority of cases the effects of seaming are to increase the crack growth rate and reduce the time-to-break of the geomembrane. Improvements to resins generally have increased the SC resistance of the geomembrane, but it does not automatically follow that similar improvements in the effects of seaming have occurred. If the fifth and sixth rows in Table 1 are compared, the crack growth rates of the geomembrane have decreased for the "new" resin but have increased at the seam. Only in the odd instance has seaming not reduced the SC resistance of the geomembrane.

Geomembrane Microstructure

To determine the factors that might cause an increase in the SC susceptibility of the geomembrane at the edge of the seam, the microstructure of a geomembrane on each side of a stress crack was investigated. Small cubes were cut out of the top surface of the bottom geomembrane from each side of an actual stress crack at the edge of an extruded seam. Therefore, some specimens would be from the "pool" region underneath the extrudate. Additional specimens were removed from the geomembrane well away from the seam for reference purposes. The following measurements were made on the small cubes:

- Density, by density gradient column, and
- Crystallinity and oxidative induction temperature (OIT), by differential scanning calorimetry.

The results of the density traverse are shown in Figure 5. The crack appears to have occurred at a point where there is a change in density, an increase from underneath the extrude out into the parent geomembrane.

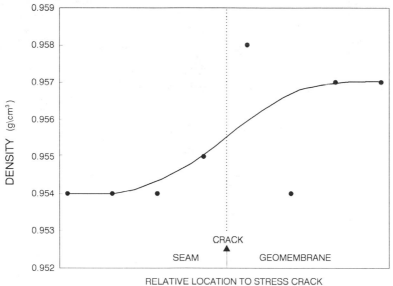

FIG. 5 -- Density distribution around a stress crack.

Figures 6 and 7 show the crystallinity and OIT traverses, respectively. There appears to be an increase, or a perturbation, in crystallinity to match the change in density, with the stress cracking occurring on the seam side of the increase. The OIT shows an increase from the geomembrane into the seam area. These preliminary results are not conclusive and need confirmation. Such investigations are ongoing, under contract to the Electric Power Research Institute, on all the available geomembranes and types of seams.

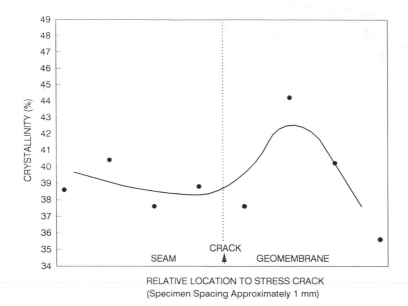

FIG. 6 -- Crystallinity (%) on the side of a stress crack.

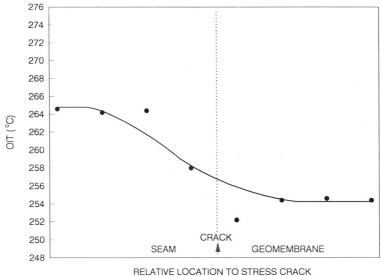

FIG. 7 -- OITs on the side of a stress crack.

Infrared spectrophotometry was performed on material removed from the fracture face and the geomembrane itself. The two spectra were identical and showed only bands attributable to PE.

While there is a change, or perturbation, in the density and crystallinity of the geomembrane in the region of the crack, it appears that the cracking may not occur at the maximum density and crystallinity but rather at the edge of the change. It is, therefore, possible that structural orientation changes at the edge of the "pool" provide a zone of weakness that accelerates the crack growth rates.

Whatever the mechanism, this information can be used practically to optimize seaming procedures and the durability of liners. In the failed installations, it is necessary to make appropriate repairs that will prevent future SC and RCP failures. In addition to installing compensation panels that will allow the geomembrane to contract without inducing high contraction stresses, it is necessary to repair all nonpenetrating stress cracks so that there is no chance they will propagate to become fully penetrating cracks in the future. A proposed repair method which must remain proprietary but which involved laying a special extruded bead over the top of stress cracks was evaluated by performing CTL tests to ensure that the pre-existing stress crack would no longer initiate premature SC failures and that the edge of the repair would not reduce the SC resistance of the geomembrane any more than did a satisfactory original seam. Figure 8 shows that three attempts were required to find the correct seaming parameters to produce a repair that adequately deactivated the original stress crack and produced SC resistance at the edge of the repair equal to the SC resistance at the edge of the original seam.

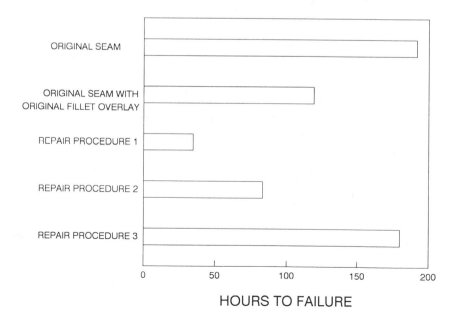

HOURS TO FAILURE

FIG. 8 -- CTL failure times of geomembrane at edges of seams and repairs.

CONCLUSIONS

This preliminary study has shown that the majority of seaming procedures cause a reduction in the stress cracking resistance of adjacent geomembranes to different degrees. There is an indication that this reduction may be caused by a combination of microstructural reorientation effects at the edge of the resolidified weld material and secondary crystallization of the geomembrane in the heat-affected zone of the adjacent geomembrane.

The CTL analytical procedures used to assess the durability and SC resistance of geomembranes can be used to optimize seaming procedures to minimize reductions in SC resistance at the edges of seams.

REFERENCES

[1] Lustiger, A., "The Molecular Mechanism of Slow Crack Growth in Polyethylene," Proceedings of the Eighth Plastic Fuel Gas Pipe Symposium, American Gas Association et al., 1983 (also other papers in the Eighth, Ninth, Tenth, and Eleventh Plastic Fuel Gas Pipe Symposia, 1983, 1985, 1987, and 1989).

[2] Peggs, I.D., "Stress Cracking of Polyethylene Geomembranes: Field Experience," Proceedings of the Second GRI Seminar -- Aging and Durability of Geosynthetics, Philadelphia, PA, December 1988.

[3] Halse, Y.H., Koerner, R.M., and Lord, A.E., Jr., "Laboratory Evaluation of Stress Cracking in HDPE Geomembrane Seams," Proceedings of the 2nd GRI Seminar -- Aging and Durability of Geosynthetics, Philadelphia, PA, December 1988.

[4] Yayla, P., and Leavers, P.S., "A New Small Scale Pipe Test for Rapid Crack Propagation," Proceedings of the Eleventh Plastic Fuel Gas Pipe Symposium, American Gas Association et al., San Francisco, CA, October 1989.

[5] Peggs, I.D., and Winfree, J.P., "Understanding and Preventing 'Shattering' Failures of Polyethylene Geomembranes," verbal presentation at the Fourth International Conference on Geotextiles and Geomembranes, International Geotextile Society, The Hague, The Netherlands, May 1990.

[6] Charron, R.M., "Polymers for Synthetic Lining Systems: Some Molecular Structure-Property-Application Relationships," Proceedings of Geosynthetics '89, San Diego, CA, February 1989.

[7] Geosynthetic Research Institute, "Ductile/Brittle Transition Time for Notched Polyethylene Specimens under Constant Stress," GRI Test Method GM 5, Revised June 23, 1989.

[8] Peggs, I.D., and Carlson, D.S., "Microtome Sections for Examining Polyethylene Geosynthetic Microstructures and Carbon Black Dispersion," Proceedings of Geosynthetics '89, San Diego, CA, February 1989.

Ernst Schmachtenberg   and   Karlheinz F. Bielefeldt

DEVELOPMENT OF A SHORT TERM TESTING METHOD FOR WELDED LINERS

REFERENCE: Schmachtenberg, E., and Bielefeldt, K., "Development of a Short Term Testing Method for Welded Liners Geosynthetic Testing for Waste Containment Applications, ASTM STP 1081, Robert M. Koerner, Ed., American Society for Testing and Materials, Philadelphia, 1990.

ABSTRACT: One of the critical areas of a sealing made of polymer liners is the seam. The difficult and often changing conditions during welding on the site are the reason for this. Since non-destructive testing methods do not control the adhesion of the junction, destructive testing of welded samples is necessary. With the control of the welding parameters during the welding process, constant quality can be guaranteed. Therefore it is of great interest, to find testing methods which allow to improve the quality of the seam in a short term testing procedure. This way, the parameters for the welding process and seam quality can be optimized on the site.

    In a reseach project, sponsored by the German Department of the Environment and performed by the Süddeutsche Kunststoff-Zentrum, several tests on welded samples were carried out, related to the different properties. To test the adhesive forces in the seam, peeling tensile tests at elevated temperatures and constant loads allow best distinction. Correlations between time to failure in the long-term peel test and geometry of the seam could be found. For correlations between ageing of the polymer material caused by the welding and internal stress in the seam, no simple test procedure could be found up to this day. The project will be continued.

KEYWORDS: polyethylene lining, welding of plastics, long-term properties, quality control, testing methods, seaming parameters, liner surface, morphology.

Dr. Schmachtenberg is chairman and Dr. Bielefeldt is research scientist at the Testing, Research and Development Division, Süddeutsches Kunststoff-Zentrum (SKZ), Frankfurter Straße 15, D-8700 Würzburg, FRG

INTRODUCTION

The goal of the investigations presented here is to develop a test procedure for evaluating the quality of welded joints of polymer liners for waste disposal facilities made of HDPE. The expectations to be met by the test procedure are, on one hand, to give an indication regarding the quality of the welded seam within a short period of time, preferably already at the construction site, on the other hand, to ensure the optimization of welding parameters and welding procedures.

In the Federal Republic of Germany methods for examining the suitability of polymer materials for waste disposal site liners are backed by comprehensive works of regulation, describing the generally acknowledged state-of-the-art technology. For the testing of the welded joints on waste disposal site liners, however, no satisfactory solution has been found so far. Today short-term peel tests are performed which at least make it possible to identify the defective welded joints: if the seam peels, the quality is unsatisfactory; if the polymer liner fails in the boundary area of the welded joint due to stretching of the base material, the welding quality is sufficient. However, the results of the short-term peel tests do not allow any further optimization of the welding techniques respectively of the process parameters for welding.

THEORY

The term welding describes the joining process, where two parts are connected with each other by melting the boundary areas, and in some cases by using additional melted material of the same kind, and joined under pressure. Ideal conditions allow joints with mechanical properties very close to those of the starting material.

The joining properties of a welded joint are based on various physical mechanisms. The strength of a welded joint is determined by the intramolecular bonding forces [1]. These are the secondary valency forces (Van der Waals Forces) which are active at a distance of approx. 10^{-9} m from the center of the molecules. In general one distinguishes between the dipole-, induction-, dispersion- and hydrogen bridge bonding forces.

The various adhesion mechanisms of material combinations can in general be derived from that. Ideal adhesion is secured whenever the surfaces to be joined are of identical material, as it is the case when welding non-oxidized or uncontaminated materials (adhesion theory).

In view of the minimal distance in which secondary valency forces are active, the viscoelastic contact theory was derived. In simple terms this theory says, that only after the surface profiles of the parts to be joined have been adapted, a sufficient flow of force is guaranteed. This approach demonstrates why a specific pressure per unit area is required for a joining process.

This model is still further extended in the diffusion theory where the assumption prevails that a diffusion of chain segments takes place in the boundary area. In this case the primary valency bonds of the macromolecules contribute in building up the weld strength. Due to the high viscosity of plastic melts, however, the diffusion rate is low. Only when a flow is forced upon, similar to what happens in an extruder during processing, a complete diffusion of the chain segments beyond the boundary areas is ensured.

This "interpenetration" of molecule segments promotes physical entanglements, the boundary areas completely disappear optically as well as mechanically. The further the chain segments diffuse into the boundary area and the more entanglements emerge, the better the properties of the welded joint are. This process is, however, directly dependant on the welding parameters. Relevant is the attainable flow rate of the melt [2], which is determined by the process parameters temperature, pressure and time. However, an optimum is to be expected here, since transverse orientations influencing the strength may develop when the melt flow is too high. Further, there is a risk that molecule segments having just entangled are being forced out of the weld seam region.

On the other hand, by squeezing the material, contaminations, especially decomposed products and dirt from the liner surface are removed from the seam   [3, 4].

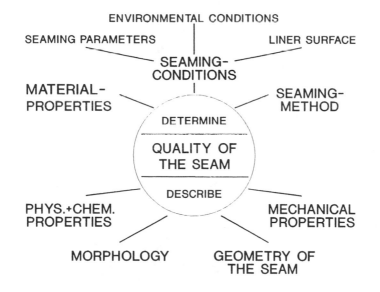

Fig. 1: Quality of the Seam

Overall the welding technique, the hereby selected parameters of the joining process (temperature, pressure, time) and the used material determine the quality of the seam (Fig. 1). Besides the mechanical properties further characteristics of quality are morphology, the physical and chemical properties of the material in the seam area and the geometry of the seam, produced by the welding process.

EXPERIMENT

In a slit die extrusion line polymer liners (2 and 3 mm thick) made of carbon-black-filled polyethylene of high density were extruded to serve as test specimen. The welding operations were partly carried out by a hot gas welding unit and partly by heated wedge welding units. With both procedures overlapping double welded joints were produced. The parameters temperature, contact pressure and velocity were varied over a wide range whereby deliberate attempts to obtain poor welded joints were made. Table 1 shows the parameters for the test specimen production:

Welding Technique	Test Sample Series	Liner Thickness (mm)	Welding Parameters			Surface Treatment
			Temperat. (°C)	Velocity (m/min)	Joining Pressure (N)	
Hot Air	1	2	450	1,0	erhöht	none
	2		470	1,0		grinded
	3		460	1,1		none
	4		450	1,6		water spray
	5		460	1,0		none
	6		450	1,0		"
	7		360	1,0		"
	8		610	1,0		"
	9		600	1,8		"
	10		600	1,9		"
Hot Wedge	21	2	360	1,2	780	dry,
	22		350	2,6	800	cleaned
	23		340	1,2	1190	with cloth;
	24		350	1,2	770	water spray
	25		360	0,8	780	dry,
	26		460	1,2	710	cleaned
	27		280	1,2	800	with cloth
	29	3	330	1,2	700	"
	30		330	1,2	1400	"
	31		470	1,2	1400	"
	32		270	1,2	780	"
	33		380	1,2	700	"
	34		380	1,5	770	"

Table 1: Production of Weld Test Samples

The analysis of the weld contour and the bead was of particular interest for the evaluation of the weld quality. In order to make the weld visible, test strips were cut out of the polymer liner at a right angle to the weld joint. The strips were then annealed for 6 hours at 100 °C. This annealing process retracts the internal stresses brought about by the welding operation which in turn leads to the deformation of the cut surface. The cross sectional area melted during the welding operation assumes a concave shape (Fig. 2). In this manner the contour of the weld can now be measured simply with the aid of a workshop microscope.

After preliminary tests the following test methods were chosen for the mechanical examinations of the weld joints:

Weld Line Tests	Temperature (°C)	Clamping Distance (mm)	Test Result
1. Short-Term Peel Test (Test Samples No. 5 according to DIN 53 455) v = 50 mm/min	23	40	Peel Resistance (N/mm)
2. Tensile Impact Test (Test Samples B, following DIN 53 448)	23	30	Tensile Impact Strength (mJ/mm^2)
3. Long-Term Peel Test (Test Samples like in No. 1, Long-Term Line Load 6 N/mm)	70	~ 40	Time to Failure, Interruption after 1000 Hours (h)
4. Long-Term Shear Test (Test Samples like in No. 1, Long-Term Stress)	80	130	Time to Failure, Interruption after 1000 Hours (h)

Table 2: Mechanical Testing at Welded Seams

Representative of the short-term peel tests are the results of hot air welded test samples illustrated in Figure 3. Distinctly defective welded joints, like samples 3, where the surface was treated with a tenside incorrectly before welding, can be detected with this method. Though it is not possible to make a distinction between more or less good seams.

Tensile impact tests were made (Fig. 4) in order to detect possible weaknesses in the transition zone from the welded joint to base material. Results show that the impact energy evidently has different effects on the various weldings, although a systematic correlation of the bead cannot be recognized. It is striking, however, that for particularly defective welded joints (e.g. test samples 3) some of the best tensile impact strengths are determined.

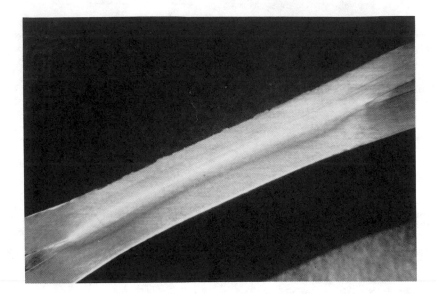

Fig. 2: Cross Section of a Seam after Annealing

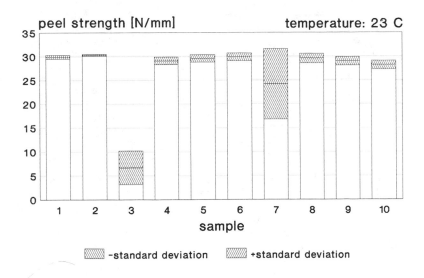

Fig. 3: Tensile Peel Test (Hot Air Welded Samples)

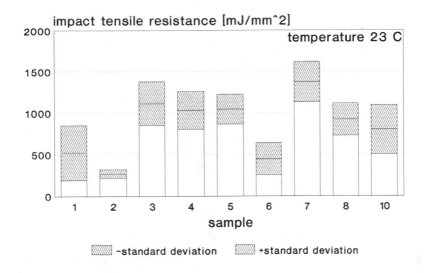

Fig. 4:   Impact Tensile Test (Hot Air Welded Samples)

The long-term peel test supplies far better information about the quality of welded joints (e.g. Fig. 5 for 2 mm thick hot wedge welded liners). The test submits a relatively refined estimate of the various test samples.

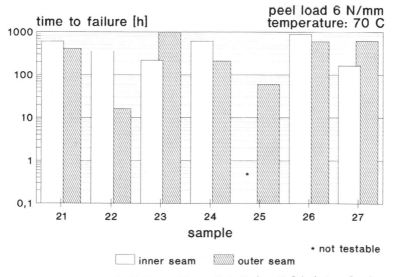

Fig. 5:   Long-Term Peel Test (2 mm Hot Wedge Welded Samples)

Not only defective (e.g. 22, 25) but also good seams (e.g. 23, 26) can be identified. Contrary to the short-term tests, this method applies exclusively to the welded joint (failure occurs in the base material). With the exception of samples where the test was interrupted after 1000 hours, all pieces failed due to peeling of one or both welded joints (Fig. 6).

Fig. 6:   Failure Mode in the Long-Term Peel Test (Hot Wedge Welded samples)

The long-term shear tests carried out by us prove to be less suited for examining weld seams (Fig. 7), since they merely provoke a failure in the base material. The failure occurs, independent of the liner thickness, immediately next to the welded joints (Fig. 8).

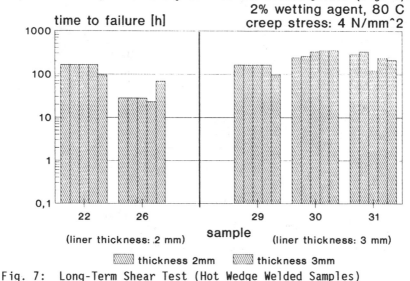

Fig. 7:   Long-Term Shear Test (Hot Wedge Welded Samples)

Fig. 8:   Failure Mode in the Long-Term Shear Test (Hot Wedge Welded
          Samples)

Although a certain correlation with the process parameters can be
observed. Close observation shows, the time to failure is influenced
negatively by a relatively high welding pressure as in samples 26 or
by an apparently too low pressure such as in samples 29.

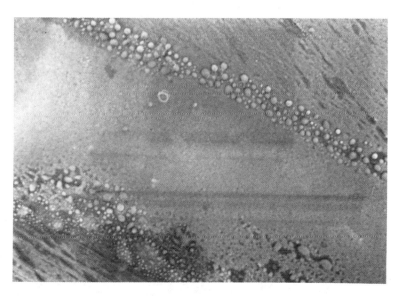

Fig. 9:   Contamination in the Bead

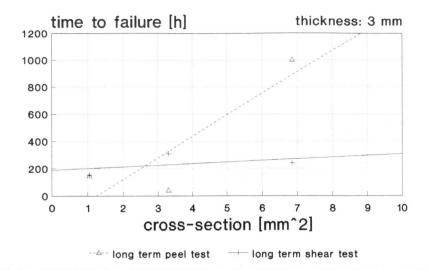

Fig. 10: Time to Failure as a Function of the Cross Section of the Bead

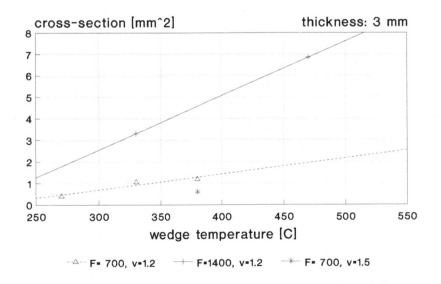

Fig. 11: Cross Section of the Seaming Bead as a Function of Welding Parameters

The theoretical considerations made earlier already touched on the importance of the bead. Microscopical investigations (e.g. Fig. 9) show that contamination is pressed from the seam into the bead. This contamination may be decomposed products, agglomerated carbon black particles or dirt on the surface of the polymer liner.

The cross sectional area of the bead is significant for the flow process during the welding operation. First test results (trend analysis) indicate that the time to failure in the long-term peel test can be related to the bead area (Fig. 10). The time to failure of welds on both 2 mm and 3 mm thick PE-liners increases as the bead grows. The long-term shear test, in contrast, does not indicate a clear correlation between the bead area and the time to failure.

A relationship between the cross sectional area of the bead and the process parameters was observed in the case of hot wedge welding. Figure 11 shows the cross sectional area of the bead polymer liners, 3 mm thick and welded with the hot wedge, as a function of the welding parameters. As the temperature of the hot wedge rises and the joining force increases, the higher speed of advance of the welding unit (the other parameters remaining the same) leads, as expected, to a reduction of the cross sectional area of the bead, which in turn is directly related to the reduced local melting rate, due to the shortened contact time. Insofar it makes sense, that the size of the cross sectional area of the bead is a measure for the combination of the welding parameters temperature, joining force and welding rate determining the flow process.

CONCLUSION

1. Defective seams can be detected in short-term peel tests. But this method can not be used to differentiate between sufficient and good quality. The influence of the liner thickness has to be regarded [5].

2. It was expected that the results from the tensile impact tests on lapped seams would provide information on the notch influence in the seam transition area. So far, however, no systematic correlation between the welding parameters and the results from the tensile impact test could be found.

3. The long-term peel test at 70 $^\circ$C and a peel load of 6 N/mm proves to be good to evaluate the seam quality. With this test method not only poor seams can be detected but it is possible to differentiate between sufficient and good quality. Further investigations must now show to what extent this procedure can be applied to different polyethylene materials.

4. Tensile creep tests (at 80 $^\circ$C, 2 % wetting agent solution in water, 4 N/mm^2), as they are used for testing butt welded pipes, are not able to point out particularly bad seams. In this respect they do not seem adequate for testing lapped joints.

5. The size of the bead is suitable for a simple quality evaluation (e.g. at a construction site). On one hand the bead area correlates with the welding parameters, on the other hand with the results from the long-term peel test. These relationships could be verified for a series of hot wedge seams. Planned future work in this project will show, if these correlations can be applied to other welding methods.

ACKNOWLEDGEMENTS

The authors acknowledge with appreciation the help of Dr. Heino Klingenfuss, Leopold Glueck, Friedrich Kurr, Karin Mitchell, and Susanne Weiss in the preparation of this paper.

REFERENCES

[1] Haberstroh, E., Potente, H., " Fügen und Umformen von Kunststof-fen", Skriptum zur gleichnamigen Vorlesung in der überarbeiteten Fassung WS 88/89

[2] Potente, H., " Eine Analyse des Heizelementstumpfschweißens von Rohren aus teilkristallinen Thermoplasten", DVS Berichte, Band 84, Düsseldorf 1983, pp. 41-49

[3] Horn, V., " Parameterabhängige Gefüge, Festigkeit und Zähigkeit von heizelementgeschweißtem Polypropylen", Schweißtechnik, Berlin (GDR), Vol 37, No. 10, 1987, pp. 450-451

[4] Stanjavicius, V. I., Kevjalajtis, Z. A., Savickij, A. Z. S., " Examination of the Degree of Oxidation of Polyethylene at High Temperatures Welding with Heated Tools", (Titel from Russian), Svarka i skleiv izd iz polim mat. 1987, pp. 142-149

[5] Richardson, G.N., Koerner, R.M., " Geosynthetic Design Guidance for Hazardous Waste Landfill Cells and Surface Impoundments", Preliminary Draft, Office of Research and Development U.S. Environmental Protection Agency, Feb. 1987, p. EPA III - 13

Richard T. Sprague, and Ronald K. Frobel

PERFORMANCE TEST METHODS FOR GEOMEMBRANES AND GEOCOMPOSITES USED IN THE DESIGN OF WASTE CONTAINMENT FACILITIES

REFERENCE: Sprague, R. T., and Frobel, R. K., "Performance Test Methods for Geomembranes and Geocomposites Used in the Design of Waste Containment Facilities, "Geosynthetic Testing for Waste Containment Applications, ASTM STP 1081, Robert M. Koerner, editor, American Society for Testing and Materials, Philadelphia, 1990.

ABSTRACT: During the design process for today's generation of waste containment facilities, the design engineer must consider the use of geosynthetics, whether required by regulatory guidelines or sound engineering judgement. Just as with other civil engineering materials, geosynthetics are tested and evaluated using a variety of test methods.

This paper describes three performance test methods, two of which are draft test methods which are currently being used to evaluate performance properties of geomembranes and selected geocomposites:

o   Wide strip tensile test;
o   Direct shear test; and
o   Large scale hydrostatic pressure test.   '

The paper will attempt to correlate performance test methods with actual field performance based upon case histories of success and failure in geomembrane systems.

KEYWORDS: Waste containment, geomembrane, geocomposite, performance testing, test methods, wide strip tensile testing, hydrostatic testing, direct shear testing

Mr. Sprague is senior project manager at Roy F. Weston, Inc., Lakewood, CO 80228; Mr. Frobel is principal in R. K. Frobel Associates, Evergreen, CO 80439.

Geosynthetics (i.e., geomembranes, geotextiles, geocomposites, geonets, geogrids) are being utilized more and more frequently in the design of solid and hazardous waste landfills, industrial and hazardous waste impoundments, mining process and waste impoundments and municipal wastewater facilities.

The design engineer is faced with the selection of geomembranes, geotextiles or geocomposites based on a number of design parameters including tensile strength, puncture resistance, surface friction, chemical compatibility, seam strength, etc. He must rely on test data either supplied by the manufacturer or obtained in independent testing. In analyzing test data for design purposes, the design engineer must carefully consider the differences between what is commonly referred to as index tests and performance test methods.

ASTM Committee D35 on geosynthetics has developed an important distinction between index test methods and performance test methods. Index tests are defined as those test methods "which may contain a known bias, but which may be used to establish an order for a set of specimens with respect to the property of interest" (ASTM D 4885). Index test methods are commonly used in manufacturing quality control, materials screening or as a "quick" reference test method in comparing products.

Performance tests are defined as those test methods which "simulate in the laboratory as closely as practical selected conditions experienced in the field and which can be used in design" (ASTM D 4885). Performance tests are generally larger in scale than the typical index test and are generally slow, expensive and sometimes difficult to perform. They are, however, necessary to accurately predict field performance and it must be recognized that performance tests are currently being used in the design process.

Figure 1 shows an example of a waste containment facility, in this case a landfill, in which a geomembrane is being installed. The landfill is constructed with 1:1 slopes, and into a glacial till geological setting. The exposed landfill slope soils consist of interbedded sands and gravels, with discontinuous lenses of clayey sand and sandy clay; the foundation soils in the bottom of the landfill excavation consist of glacial lake bed clays. During installation, the geomembrane may experience sudden, uniaxial tensile loadings induced by wind or by earth moving equipment; however, during the active and post-closure life of the landfill, the geomembrane will more likely experience slow tensile loadings, either in an uniaxial or multiaxial mode. In addition, during the active life, the geomembrane may experience slow puncture or tearing forces. Figure 2 shows an example of a waste containment impoundment lined with a geomembrane. The geomembrane is experiencing multiaxial stresses due to differential settlement of the foundation soils. Each of these performance properties is addressed by a test method which is discussed in this paper. It is interesting to note that the geomembrane first selected for the landfill in Figure 1 failed catastropically prior to placement of waste, apparently due to performance properties (in this case, tear propagation resistance which is not discussed in this paper). Subsequently, geomembranes were selected based upon the critical performance property.

Although there are other performance type tests being used on various types of geosynthetics, this paper will examine three of the most ASTM test methods and/or ASTM draft test methods currently under development. The three methods to be examined are:

o   Wide Strip Tensile Test;
o   Direct Shear Test; and
o   Large Scale Hydrostatic Pressure Test.

An attempt has been made to examine case histories where these types of test methods have been used to simulate the actual field performance.

WIDE STRIP TENSILE TESTING

The ability of geosynthetics to withstand the slow application of tensile forces can be an important design property, especially in deep, steeply sloping landfill designs. In landfills, settlement of the foundation soils directly beneath the geomembrane, or other geosynthetic, can result in tensile stress in the geosynthetic. If the settlement is near the toe of the slope, this stress may be transferred to the geosynthetic as uniaxial stress; if the settlement is on the embankment or on the bottom of the waste containment cell, the stress may be transferred to the geosynthetic as a multiaxial stress (see a following section of this paper).

Similary, settlement of the waste within a waste containment cell can result in tensile stress on a geosynthetic used as a cover for the cell. In addition, settlement of soil covering the geosynthetic can result in tensile stress on the geosynthetic, whether that soil cover is part of a cap or part of the protective layer placed upon a geosynthetic within a landfill cell.

Although a variety of test methods have been available for testing geomembranes, geotextiles and other geosynthetics for tensile properties, none of these test methods have been designed to be large-scale performance tests. For example, ASTM Test Methods D 412 ("Standard Test Method for Rubber Properties in Tension"), D 638 ("Standard Test Method for Tensile Properties of Plastics"), and D 751 ("Standard Methods for Testing Coated Fabrics") specify speeds varying from 0.85 mm/s to 8.5 mm/s depending on the type of material undergoing testing; in addition, these test methods specify specimens of dimensions differing from 1.0 cm (dogbone) to 6.5 cm, again depending upon the material undergoing testing. These test methods were not all developed for testing of gemembranes; the test methods for geotextiles (eg., Test Method D 4632, "Standard Test Method for Breaking Load and Elongation of Geotextiles (Grab Method)") add an additional variable in terms of the clamping system. All of these test methods were developed to allow for measurement of properties important for a distinct type or family of geosynthetic, and therefore, do not allow for testing of properties important to other types of geosynthetics, and do not allow for direct comparison of different materials.

Test methods D 4595 ("Standard Test Method for Tensile Properties of Geotextiles by the Wide-Width Strip Method") and D 4885 ("Standard Test Method for Determining Performance Strength of Geomembranes by the Wide Strip Tensile Method") are recently established methods for assessing performance Tensile properties in geotextiles or geomembranes, respectively.

Summary of Method

The same basic test method is used for both geomembranes and geotextiles, with minor modifications for these two types of materials in the rate of extension (elongation). A wide specimen is gripped across its entire width in the clamps of a constant rate of extension tensile testing machine. The testing machine is operated at a prescribed rate of extension, and applies a uniaxial load to the specimen until the specimen ruptures or otherwise fails. The tensile strength, elongation, initial or secant modulii, and breaking toughness are calculated from the measuring devices on the testing machine.

The prescribed rate of extension for testing geomembranes is 0.0167 mm/s (1 mm/min); the rate of extension for geotextiles is 0.167 mm/s (10 mm/min). This slow rate of extension, especially for geomembranes, simulates as closely as practical the conditions expected in the field, making wide strip tensile a performance test.

Case Study	Wide Strip Tensile Testing
	Bayview Landfill
	South Utah Valley Solid Waste District
	Provo, Utah (HDR, 1989)

Bayview Landfill is a new facility approximately 30 miles southwest of Provo, Utah, and near the town of Elberta. The landfill site consists of 267 hectares of semidesert rangeland, and has a planned active life of at least 50 years. The first phase of the landfill includes design and construction of half of the first cell, covering approximately 9 hectare. This cell is designed to be between 10 and 13 meters deep, with 4:1 sideslopes (see Figure 3).

Conceptual design of the first landfill cell included a liner system consisting of a geomembrane between two layers of geotextile. The underlying geotextile was intended to provide protection for the geomembrane: the excavation had been completed under prior contracts, and included rocks with a maximum diameter of 10 cm in the compacted soil base. The puncture resistance requirements for this base soil could have been modelled using the large scale hydrostatic punture test discussed below; however, in this case, experience and existing literature was used to select a 285 g/m^2 geotextile. Both conceptual and final design included a 1 mm HDPE geomembrane to control leakage from the landfill cell.

Conceptual design also included a 95 g/m^2 geotextile on top of the geomembrane, with the geotextile intended to reduce the potential for puncture. The uppermost geotextile was then to be covered with 0.6 meters of soil cover consisting of gravelly, silty sand, with a maximum particle diameter of 6 cm. During final design, the upper geotextile was evaluated for its ability to perform under the design conditions. This analysis indicated that the critical performance property would be the ability to withstand uniaxial tensile loading during placement of the cover soil, and movement of waste hauling equipment on the landfill ramp. Using a factor of safety of 2.0 under dynamic loading, the analysis indicated that the geotextile would require a performance strength of 208 kg/m.

Based upon wide strip tensile test data, it was established that a 285 g/m^2 geotextile had the required strength.

This landfill was constructed during the fall of 1989.

DIRECT SHEAR PERFORMANCE TESTING

The coefficient of friction between a geosynthetic and soil or between two geosynthetics can become a critical design property, particularly for geomembrane lined slopes of waste impoundments and covers. Although not yet standardized, the draft method D 35.01.81.07, "Standard Test Method for Determining the Coefficient of Soil and Geosynthetic or Geosynthetic and Geosynthetic Friction by the Direct Shear Method" [1], is a direct adaptation of the conventional geotechnical engineering direct shear test (ASTM D 3080) used in the determination of soil shear strength. This type of test is valuable in determining site specific friction values of materials used in both the bottom containment and closure of a waste facility.

The consolidation of waste in a solid or hazardous landfill will induce movement of the waste relative to the geomembrane or geosynthetic. If the geomembrane is restrained, deformation occurs between the soil and geomembrane and as deformation progresses, increased shear stress is mobilized at the interface of soil/geomembrane. The slope stability, therefore, is directly a function of the shear strength at the interface between the soil and geomembrane or between the soil and geotextile if a geotextile is used as a protective layer. The critical interface friction angle here may be between the geomembrane and geotextile; however, the design engineer should assess all interfaces.

The lining system in a cap is used to cover the waste after landfill closure to help eliminate infiltration of rainfall and minimize leachate generation. Many cap designs are composite designs with several layers of geosynthetics and soil. As soil cover materials move down the cover slope, shear stresses are mobilized between each of the interfaces. This mobilized shear stress must be large enough to support the soil layer on a stable slope. Normally, stability is an issue on the steeper portions of the containment and cap where slopes range between 4:1 (14 deg.) and 2.5:1 (22 deg.).

A convenient method to evaluate the coefficient of friction between a geosynthetic and soil or any combinations of geosynthetics is by the direct shear method. In the ASTM draft method D 35.01.81.07, a constant normal compressive stress is applied to the specimen and a tangential (shear) face is applied to the apparatus so that the lower section of the test apparatus moves in relation to the upper section. Figure 4 is a schematic of the test apparatus. The major difference between the "performance direct shear" and standard geotechnical direct shear is the size of the test apparatus, which is required to have contact surfaces which are a minimum of 300 mm x 300 mm, and therefore the size of the specimen is required to be much larger. This is due to the fact that when soils are displaced relative to a rough surface geosynthetic, the deformation required to mobilize residual shear stress is between 20 and 80 mm; therefore, the contact area must be much larger than for conventional direct shear in soils.

Summary of Method

A constant normal compressive stress is applied to the specimen while a shear force is applied to the lower section of the test apparatus. The shear force is recorded as a function of the deflection of the moving section of the shear box. The test is performed at a minimum of three different normal stresses representative of the field conditions. The peak (or residual) shear stresses are recorded and plotted against the normal compressive stresses used for testing. Test data is modeled by a best fit straight line whose slope is the coefficient of friction between the two materials being tested. The y-intercept is the adhesion value. Figure 5 illustrates a typical failure envelope based on three different normal compressive stress conditions.

This type of test is a performance test in that it can provide the designer with values for the "site specific" soils and geosynthetics proposed for use in a given project.

Case History   **Direct Shear Performance Testing**
                                     San Justo Reservoir
                                     Central Valley Project
                                     California (USBR [2])

During the summer and fall of 1985 a 1 mm thick HDPE geomembrane was installed over portions of the San Justo Reservoir during construction of embankments at the site. The geomembrane was necessary to prevent excessive reservoir seepage in areas that were prone to landslide outside the reservoir area. The geomembrane was covered by a layered system of earth materials. During February 1986, after unusually heavy rainfall, approximately 22 percent of the lined area experienced slope failure wherein the cover materials slid down slope and in some cases exposed the geomembrane (Figure 6). Slopes varied between 4:1 and 2.5:1.

A laboratory study was undertaken by the U.S. Bureau of Reclamation to determine the cause of the slope failures. The laboratory study consisted of a series of Direct Shear Performance Tests using the cover materials and the HDPE lining to evaluate the coefficient of friction of the lining and to conduct stability analyses.

Three different soil types (a sandy clay, a poorly graded sand and a fine sand) were tested against the original HDPE and a variety of surface treatments and new liner materials designed to increase the surface friction. Soil materials were tested in saturated conditions. The types of geomembrane surfaces tested were as follows:

Smooth	Original HDPE geomembranes
Sandblasted	Original HDPE geomembrane roughened by sandblasting
Wire brush	Original HDPE geomembrane roughened by stiff wire brush

Geonet	Geonet attached to the original HDPE geomembrane
Texturized	1.5 mm thick roughened or texturized HDPE geomembrane (Gundle Lining Systems)
Embossed	Embossed HDPE geomembrane (Schlegel Lining Technologies)

The Direct Shear Testing provided site specific data that enabled the design engineers to determine the cause of failure and the remedial actions necessary to treat the exposed areas of the HDPE geomebrane. It was determined that sandblasting would sufficiently increase surface friction on most slopes to allow replacement of a 0.15 m thick layer of free draining bedding topped by a 0.3 m thick layer of rock fragments. One area was critical enough to require removal of the existing liner and to replace it with a new embossed HDPE.

LARGE SCALE HYDROSTATIC MULTIAXIAL BURST AND PUNCTURE TESTING

There are currently several index test methods devoted to puncture resistance of geosynthetics. However, these tests are indeed only index tests and are applied to a small specimen area. ASTM Committee D35 on Geosynthetics is currently developing a two-part draft standard devoted to large scale hydrostatic puncture and multiaxial stress-strain performance testing. Both test methods can be performed on the same apparatus and current draft designation is D 35.10.88.01, entitled "Standard Performance Test Method for Large Scale Pressure Testing of Geosynthetics."

The test method is designed as a performance test in that it will determine the ability of a geosynthetic to conform to an irregular surface under simulated loading. Such "surfaces" or subgrades as various aggregate sizes or machined points (cones or pyramids) can be used to simulate field conditions. The specimen test area is at least 500 mm in diameter which will allow the geosynthetic to stress multiaxially over point loads or over a void area.

Large diameter multiaxial stress-strain testing on geomembranes and geotextiles has been carried out on a research basis for over 13 years. Several researchers, including Rigo [4], Frobel [5, 6], Fayoux and Loudier [7], Loudiere and Pignon [8], Frobel et al [9], and Laine et al [10], have described the various testing devices developed to simulate an in-service quantitative puncture performance test of primarily geomembranes or geomembrane/geotextile combinations. In principle, each of the devices simulate overburden or water pressure acting on a geomembrane when placed on a subgrade containing either natural aggregate or machined puncture points, to evaluate the geomembrane's ability to conform to subgrade (or cover) irregularities such as aggregate, rock out croppings, waste debris, etc. Figure 7 illustrates the mechanism of deformation according to Rigo [3]. There are two critical zones in the geomembrane - zone 1 extends over the geomembrane - aggregate contact and zone 2 comprises the void areas formed by the geomembrane between aggregates. Thus, there are two types of failure possible - rupture by puncturing (zone 1) or rupture by bursting (zone 2). Figure 8 is a photograph of the type of test vessel used for hydrostatic point load/puncture testing.

Large diameter multiaxial stress-strain geosynthetic testing has been in use in several countries including France, Germany, Belgium, the Netherlands, Canada and the United States. Diameters of pressure vessels and frames have varied between 150 mm and 1000 mm and strain rates have varied from 0.23 kPa/min to 100 kPa/min for testing on geomembranes and composites. Based on research and comparative results, a diameter of 500 mm is currently being recommended as minimum. Both working group AK 14 of the DEGEG (Federal Republic of Germany) and ASTM Committee D 35 on Geosynthetics have recommended a 500 mm diameter in their draft procedures for multiaxial burst testing. Strain rates (pressure increments) are commonly in the range of 5 to 10 kPa/min with consideration given to material creep and subsequent strain rate adjustment.

The multiaxial stress-strain test is designed to evaluate a material's response to out-of-plane loading as would occur during uneven settlement of supporting soils. This type of loading condition frequently occurs under geomembranes used in waste cell covers and over compressible or collapsible subgrade conditions. For example, a localized failure of a landfill drainage pipe under a geomembrane could cause a subgrade collapse, thus straining the geomembrane. Geomembrane seams can also be tested in this type of device.

Figure 9 illustrates a multiaxial stress-strain test device that uses upward air or water pressure to stress the membrane to rupture. This type of device is recommended for multiaxial stress-strain testing primarily due to the visual observation of the test and the ability to mark the specimen with grids, place strain gages and photograph the specimen at various stages of strain.

Summary of Method

Hydrostatic Point Loading: A large test specimen (at least 600 mm in diameter) is placed over the base of a large hydrostatic pressure vessel (minimum inside diameter of 500 mm) containing a grouping of standard puncture points. The top of the vessel is placed over the specimen, bolted and sealed in place, filled to a shallow depth with water and then pressurized gradually until rupture 2of the geomembrane occurs. Rate of pressure increase (strain rate) and puncture points can be adjusted to actual or simulated site conditions or loadings. The vessel must be designed for accurate pressure control and measurement as well as immediate failure detection.

Multiaxial Stress-Strain: A test specimen (at least 600 mm in diameter) is placed flat over the base of the vessel. The top of the vessel (or ring for upward burst) is placed over the specimen and bolted in place. The vertical deflection rod is set to 0 and initialization back pressure (approximately 1380 Pa) is applied. Pressure is increased in a step function or constant rate until failure occurs. Incremental readings are taken with time and pressure to establish an accurate stress-strain curve. Calculations for stress and strain at any given pressure and deflection are based on conventional spherical trigonometry and shell stress theory, and are presented in the ASTM draft standard. Geomembrane seams can also be tested in this fashion.

<u>Case History</u>   **Large Scale Hydrostatic Point Stress**
                                     Mt. Elbert Forebay Reservoir
                                     Fryingpan-Arkansas Project
                                     Colorado (USBR [11)

During the summer of 1980 the USBR installed 117 hectares of 1.14 mm thick CPER (reinforced chlorinated polyethylene) geomembrane in the Mt. Elbert Forebay Reservoir, near Leadville, Colorado. The Forebay is the upper regulating reservoir for the Mt. Elbert Pumped-Storage Power Plant. At the time of its construction, the reservoir constituted the world's largest single-cell geomembrane lining. Due to the critical nature of the project and the severe climatic conditions, an intensive effort was made to investigate all critical requirements of a geomembrane lining for long-term service under unique and severe operating conditions. As part of the intensive laboratory investigation into geomembrane materials and the subsequent quality assurance program, a special laboratory study was conducted using large-scale (600 mm diameter) hydrostatic pressure vessels. The geomembrane was tested over simulated subgrade conditions using loose surface aggregate similar to that found in the Forebay. The 1.14 mm thick CPER (chlorosulfonated polyethylene, reinforced) geomembrane with seams was subjected to the hydrostatic pressure at maximum design depth (35 feet) of the reservoir for two months. The pressure was raised incrementally at a slow rate to simulate filling of the Forebay Reservoir.

Although the sheet material did not physically fail, several observations on its performance were noted and subsequently used in the subgrade preparation and quality assurance (QA) program:

1.   The CPER geomembrane conformed to the subgrade irregularities within the limitations of material strain.

2.   Sharp, angular aggregate tended to separate the CPE from the scrim reinforced grid at point stresses.

3.   In laboratory testing, the subgrade under the factory seam was dry. The subgrade under the field seam was noticeably moist but no rupture in the seam was evident.

4.   The CPE had a tendency to "flow" into the grid pattern of the scrim over voids or weak spots in the subgrade.

Figure 10 illustrates the deformation that can occur in a geomembrane when tested over subgrade conditions encountered at Mt. Elbert. This hydrostatic point stress testing provided valuable information on the field performance of the geomembrane and was truly a performance test.

Figure 1.   Example of a Reinforced Geomembrane Experiencing Large Scale
Tensile and Puncture Loadings

Figure 2.   Example of a Geomembrane Experiencing Multiaxial Tensile Loadings

Figure 3.   Bayview Landfill, Provo, Utah

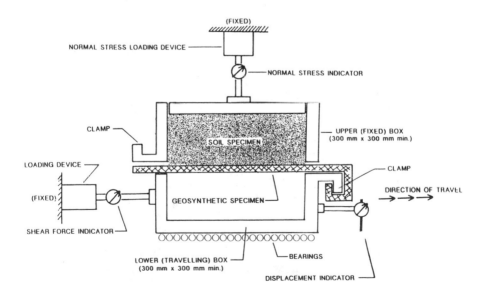

Figure 4.   Schematic of Typical Direct Shear Friction Apparatus
(ASTM draft method D35.01.81.07)

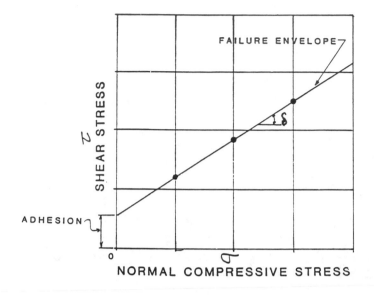

Figure 5.   Typical Failure Envelope for Direct Shear Friction Testing
(ASTM Draft Method D35.01.81.07)

Figure 6.   Slope Failure at the Soil/Geomembrane Interface, San Justo
Reservoir

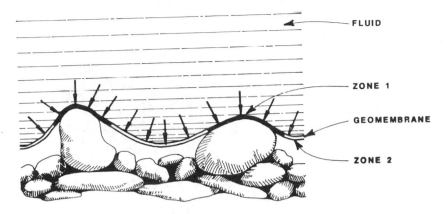

Figure 7. Mechanism of Geomembrane Deformation over Point Loadings
(adapted from Rigo [3])

Figure 8. Typical Hydrostatic Pressure Vessel Used in Geosynthetic
Conformity/Puncture Resistance (Vessel Top Removed)

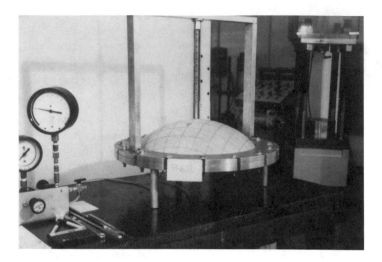

Figure 9.   Test Device Used for Multiaxial Stress-Strain Performance Testing - Upward Burst (R.K. Frobel Associates)

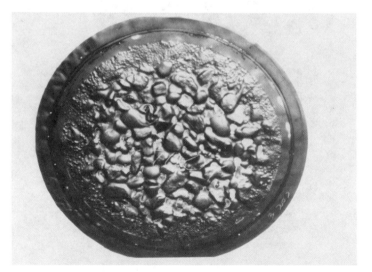

Figure 10.   Resultant Deformation of a Geomembrane after Testing Over a Simulated Subgrade

REFERENCES

[1] American Society for Testing and Materials (1989), Standard Test Method for Determining the Coefficient of Soil and Geosynthetic or Geosynthetic and Geosynthetic Friction by the Direct Shear Method - Draft Method D 35.01.81.07. ASTM Committee D35, ASTM, 1989.

[2] USBR, 1986, Membrane Liner Instability and Remedial Modifications, San Justo Reservoir, California - Technical Memorandum No. JU-230-23, U.S. Bureau of Reclamation, 1986.

[3] American Society for Testing and Materials (1989), Recommended Standard Practice for Large Scale Hydrostatic Pressure Testing of Geosynthetics - Draft Method D 35.10.88.01. ASTM Committee D35, ASTM, 1989.

[4] Rigo, J.M. (1977), Correlation of Puncture Resistance Over Ballast and the Mechancial Properties of Impermeable Membranes - Materiaux et Constructions, Vol. II, No. 65, RILEM, 1977.

[5] Frobel, R.K. (1981), Design and Development of an Automated Hydrostatic Flexible Membrane Test Facility - REC-ERC-80-9, U.S. Bureau of Reclamation, Denver, Colorado, 1981.

[6] Frobel, R.K. (1983), A Micro Computer-Based Test Facility for Hydrostatic Stress Testing of Flexible Membrane Linings - Proceedings Colloque sur L'Etancheite Superficielle Des Basins, Barrages at Canaux, Paris, 1983.

[7] Fayoux, D. and D. Loudiers (1984), The Behavior of Geomembranes in Relation to the Soil, - Proceedings International Conference on Geomembrane, Denver, IFAI Publishers, 1984.

[8] Loudier, D. and N. Pignon (1983), Puncture Resistance of Geomembrane - Proceedings of the International Conference on Surface Waterproofing of Reservoirs, Dams and Canals, Paris, February 1983.

[9] Frobel, R.K., W. Youngblood and J. Vanderoort (1987), The Composite Advantage in the Mechanical Protection of Polyethylene Geomembranes - Proceedings Geosynthetics 87, New Orleans, IFAI Publishers, February 1987.

[10] Laine, D.L., M.P. Miklas and C.H. Parr (1989), Loading Point Puncturability Analysis of Geosynthetic Liner Materials - Proceedings Geosynthetics 89, San Diego, IFAI Publishers, 1989.

[11] USBR, 1981, Installation of Flexible Membrane Lining in Mt. Elbert Forebay Reservoir -REC-ERC-82-2, U.S. Bureau of Reclamation, September 1981.

Robert M. Koerner, George R. Koerner and Bao-Lin Hwu

THREE DIMENSIONAL, AXI-SYMMETRIC GEOMEMBRANE TENSION TEST

REFERENCE: Koerner, R. M., Koerner, G. R. and Hwu, B.-L.,
"Three Dimensional, Axi-Symmetric Geomembrane Tension Test,"
Geosynthetic Testing for Waste Containment Applications, ASTM
STP 1081, Robert M. Koerner, Ed., American Society for Testing
and Materials, Philadelphia, 1990.

ABSTRACT: Presented in this paper are details of a three
dimensional, axi-symmetric, tension test used to determine the
stress-vs-strain behavior of geomembranes. From this data the
modulus, strength and elongation can be obtained. After
evaluating the effect of load rate, the accuracy of the test
was illustrated by a series of 10 tests on VLDPE geomembranes.
This test series resulted in a standard deviation of
approximately ± 10%. A variety of different HDPE geomembranes
were evaluated with only nominal differences resulting from
different manufacturing processes. Lastly, a variety of
different geomembrane types were tested and compared to one
another. All of these test results show a marked contrast to
the usually performed one dimensional tension tests and result
in substantially different behavioral trends.
    The test details and procedures are presented along with
the theory that is required to convert the pressure-vs-
deflection readings into stress-vs-strain values. The test is
recommended for the type of out-of-plane loading conditions
that occur at the bottom of reservoirs and landfills, and for
liners used for covers over subsiding landfills and other
compressible subgrade materials.

KEYWORDS: tension test, out-of-plane test, geomembranes,
axi-symmetric test, performance test, hydrostatic pressure
test, stress-vs-strain test

    Dr. Koerner, Mr. Koerner and Mr. Hwu are respectively, Director,
Senior Research Specialist and Graduate Research Assistant at the
Geosynthetic Research Institute, Drexel University, Philadelphia, PA
19104.

INTRODUCTION

One dimensional tensile tests have routinely been performed on geomembranes of different shapes and under a variety of conditions. Typical test procedures are covered under ASTM Test Methods D412, D638, D882 and D4885. All of these test methods place the geomembrane test specimen in the grips of a testing machine and apply load in a one-dimensional tension testing mode. These tests have been the "workhorse" of the industry and have considerable merit and value. Their use in the following situations is warranted and should indeed continue;

- for quality control during manufacturing,
- for comparison of processing variations,
- for comparison between different products,
- for identification purposes,
- after incubation to assess mechanical changes, and
- after aging to assess mechanical changes.

However, it should be recognized that they are *"index"* tests, and that their use in design must be done with caution.

Alternative to the use of such one-dimensional index tests for design is the use of test methods which better simulate the actual behavior in the field, i.e. *"performance"* tests. For example, the runout and anchor trench termination of geomembrane liners used for reservoirs and landfills exhibits a laterally confined, tension state which is reasonably modeled by a two-dimensional, or cross, test. The wide-width test attempts to simulate this behavior but should really be laterally confined for accurate modeling. Another type of tension stress state is that of the liner at the bottom of reservoirs or landfills, or the cover geomembrane over subsiding landfills and other compressible materials where out-of-plane tension occurs. This condition is best simulated by a three dimension, axi-symmetric tension test which is the subject of this paper.

BACKGROUND

The three dimensional, axi-symmetric tension test has been noted in the literature in a variety of forms. Frobel [1] at the U.S. Bureau of Reclamation and Laine, et al. [2] Southwest Research Institute have used large pressure vessels with geomembranes on different types of subgrade (sand, gravel, cones and pyramids, etc.) in the form of a large scale puncture test. Steffen [3], in a similar type of apparatus, gradually removed the subgrade support soil thereby generating out-of-plane tension in the geomembrane. Subsequently, he used the pressure vessel with no support soil beneath the geomembrane in the first reported three dimensional, axi-symmetric tension test. Steffen analyzed a number of different polymeric geomembranes. Subsequently, Hoechstra [4] evaluated a series of bituminous geomembranes and presented some of the requisite theory. STUVA [5] has further refined the experimental apparatus leading to the present status of this type of testing. Van Zanten [6] has also recently reported on the test illustrating the behavior of a number of geomembrane types.

In all of these efforts, the geomembrane is supported in a large pressure vessel and hydrostatically stressed until failure occurs. The resisting stresses are entirely mobilized around the circumference of the geomembrane, which conveniently forms its own seal between the two mated sections of the pressure vessel. During gradual deformation of the geomembrane, its center point deflection is monitored, thus providing deflection readings to go along with the mobilizing pressure readings until eventual failure of the test specimen occurs.

THEORY

In order to generate stress-vs-strain data from pressure-vs-deflection information, several assumptions are required as to the gradually deforming shape of the geomembrane. Using Steffen's [3] observations, in addition to our own, the following shape assumptions are proposed;

(a) at $\delta < L/2$, the shape of the geomembrane is that of a spheroid, and

(b) at $\delta \geq L/2$, the shape of the geomembrane becomes that of an ellipsoid.

In the above, "δ" is the centerpoint deflection and "$L/2$" is the radius of the pressure vessel. Furthermore, the center of the spheroid is assumed to move down along the central axis until it finally reaches the original elevation of the geomembrane where it then becomes stationary. At that time it becomes the focal point of the ellipsoid until the conclusion of the test. It should be noted that the theory could also be based on the shape of a paraboloid. Differences in strain calculated by these two assumptions, however, are quite small.

Using these shape assumptions, along with a number of trignometric relationships, one arrives at a series of equations for the desired values of stress and strain. The required formulas follow; however, the entire derivation is given in the Appendix of this paper. [7]

(a) For centerpoint deflections less than the radius (i.e., $\delta < L/2$)

• The strain calculations proceed as follows:

$$• \ R \quad = \quad \frac{L^2 + 4\delta^2}{8\delta} \tag{1}$$

$$• \ \theta \quad = \quad 2 \ \tan^{-1} \ \frac{4(L)\delta}{L^2 - 4\delta^2} \quad \text{(with } \theta \text{ in radians)} \tag{2}$$

$$• \ AB \quad = \quad R \cdot \theta \quad \text{(with } \theta \text{ in radians)} \tag{3}$$

$$• \ \varepsilon(\%) \quad = \quad \frac{AB - L}{L} \ (100) \tag{4}$$

• The stress calculations use the thickness of the geomembrane "t" and the hydrostatic pressure "p" (along with θ as defined above) to arrive at the following equation for tensile stress:

$$\bullet \ \sigma = \frac{(L^2 + 4\delta^2)p}{4(L)t \ \sin \ (\theta/2)} \tag{5}$$

(b) For centerpoint deflections greater or equal to the radius, i.e., $\delta \geq L/2$

• The strain calculations proceed as follows:

$$AB \quad = \pi \sqrt{\frac{L^2 + 4\delta^2}{8}} \tag{6}$$

$$\varepsilon(\%) \quad = \frac{AB - L}{L} \ (100) \tag{7}$$

• The accompanying tensile stress calculation is simply:

$$\sigma = \frac{\delta p}{t} \tag{8}$$

(c) The definitions for the various values in the above equations are the following:

ε = the tensile strain (in percent)
σ = the tensile stress
p = the applied hydrostatic pressure
δ = the centerpoint deflection
t = the geomembrane thickness
L = the diameter of the pressure vessel
R = the radius of the spheroidal arc (note that it is not the pressure vessel's radius except when $\delta = L/2$)
θ = the central angle of the spheroid
AB = the arc length of the spheroid

PROCEDURE

The pressure vessel used in this study is shown in Figure 1. The hydrostatic pressure is applied from above the geomembrane forcing deformation to occur in the empty lower portion of the vessel. Alternatively, pressurization could be achieved from beneath the geomembrane forcing it into an empty upper cavity. The GRI system has a diameter of 61 cm (24 in.) and is capable of sustaining 2 MPa (300 lb/in^2) internal pressure. In setting up the test, the geomembrane specimen is draped over the base, the pressure vessel lid is sealed above it, and water is introduced until the lid is full. The centerpoint measuring stick is initialized, the air vent in the upper lid is closed, the vent in the lower base is opened, and the test is ready to commence.

Water is introduced above the geomembrane at a constant flow rate thereby mobilizing the gradually increasing hydrostatic pressure. The geomembrane deforms downward into the empty base of the pressure vessel while hydrostatic gage pressure and centerpoint deflection are regularly measured. From these data corresponding stress and strain values are calculated as was described in the previous section. After failure occurs, which is signaled by a abrupt decrease in pressure accompanied by a dramatic noise, the test is complete. The pressure vessel lid is then removed and the mode of failure of the geomembrane is observed.

GEOMEMBRANES EVALUATED AND TEST RESULTS

The test sequences evaluated in the course of this study are placed in four separate groups:

- A set of 7 high density polyethylene (HDPE) geomembranes from the same roll to note the effect of varying hydrostatic pressure rates on the test results.
- A set of 10 very low density polyethylene (VLDPE) geomembranes from the same roll which was done in order to assess statistical variation of the test results.
- A set of 4 HDPE geomembranes from different manufacturers which was done in order to note variation in manufacturing processes within a single geomembrane type.
- A set of 5 different geomembranes [polyvinyl chloride (PVC), chlorosulfonated polyethylene -- reinforced (CSPE), linear medium density polyethylene (LMDPE), high density polyethylene (HDPE), and very low density polyethylene (VLDPE)] which was done in order to obtain perspective regarding a wide range of geomembrane products.

Effect of Hydrostatic Pressure Rate

In order to assess the influence of hydrostatic pressure rate on the response of the geomembrane test specimens, a series of seven tests using 1.0 mm thick HDPE were evaluated. Input pressure varied from a very slow rate of 19 l/hr. to a very rapid rate of 379 l/hr. The response curves in units of stress-vs-strain are shown in Figure 2(a) where a slight, but noticeable trend in behavior is seen. The yield strength and accompanying strain values from these curves were then plotted on Figure 2(b) where an intermediate flow rate value of 100 l/hr was selected for use in all subsequent testing.

Assessment of Precision of Test

The stress-vs-strain results of a series of 10 VLDPE geomembranes taken from the same roll is given in Figure 3. The upper set of curves show the initial portion of the response where a modulus can be seen to vary between 7 MPa and 11 MPa. While this might appear to be a large scatter, it should be noted that modulus is a very difficult parameter to monitor since considerable subjectivity is required in its determination. The lower set of curves show the entire stress-vs-strain curves where the variation in yield strength is seen to vary from 8.0 MPa to 8.5 MPa, and in yield elongation from 9% to

10%. While we do not know the precision of other types of geomembrane tension tests, this variation does not appear to be excessive and is possibly better than one dimensional test method variations.

Variation Within Different HDPE Geomembranes

The third series of tests evaluated HDPE geomembranes from four different manufacturers, see Figure 4. Here it is seen that there are differences but the general behavioral trends are similar to one another. Of particular note is that the well defined yield strength, drop off, and long plastic deformation common to one-dimensional dumbbell type tests is not present. Clearly, the manner in which the geomembrane accepts stress, hence the mode of failure (and the entire behavior leading to that point), is different.

Variation Between Different Types of Geomembranes

The fourth series of tests investigated a number of different types of geomembranes and are shown collectively in Figure 5. Their identification is as follows:

- 1.0 mm (40 mil) thick HDPE
- 0.53 mm (21 mil) thick LLDPE
- 0.76 mm (30 mil) thick VLDPE
- 0.91 mm (36 mil) thick CSPE-R
- 0.51 mm (20 mil) thick PVC

Obviously very different response curves result, as indeed they should. The point to be made, however, is that each geomembrane type has its own "signature" and for most cases it is very different from results obtained by means of conventional one-dimensional tests.

SUMMARY AND CONCLUSIONS

Clearly, out-of-plane stressing of geomembranes occurs whenever settlement, subsidence, or support material is lost beneath the liner. As such, it should be appropriately modeled in a laboratory test which simulates the actual phenomenon. The three dimensional, axi-symmetric, tension test presented in this paper is felt to be a reasonably accurate representation of such field behavior of geomembranes.

Although the test apparatus itself is quite large and bulky, the tests are easy to set up and rapid to perform (complete turn around time per test is about 2 hours). The requisite theory to convert pressure-vs-deflection readings into stress-vs-strain values has been presented. It is based on observed shapes of the deformed geomembrane under test and results in, what we feel, are reasonable results.

In addition to the test itself, it is most interesting to note the difference in stress-vs-strain behavior of conventional one-dimensional tests with the results of these three-dimensional tests on the same geomembrane materials. It is felt that this type of test is the necessary performance test for out-of-plane stressing of geomembranes.

There are many variations of the test which can be investigated and performed. The rate of application of pressure is an important decision and was not fully investigated. Extremely slow values of pressurization will allow the polymer an opportunity to stress relax and should be investigated. In a similar manner one could build pressure to a given level and sustain it, thus subjecting the test specimen to a creep-mode of stress. This variation would also be of interest to investigate. The test apparatus can reasonaby accommodate seamed specimens to note their behavior under out-of-plane stressing. Lastly, temperature variations can be examined by placing the pressure vessel in a suitable environmental chamber.

REFERENCES

[1] Frobel, R. K., "Design and Development of an Automated Hydrostatic Flexible Membrane Test Facility," REC-ERC-80-9, U.S. Dept. of Int., Bu. Rec., 1981, 25 pgs.

[2] Laine, D. L., Miklas, M. P. and Parr, C. H. (SWRI), "Loading Point Puncturability Analysis of Geosynthetic Liner Materials," Proc. Geosynthetics '89, San Diego, CA, IFAI, pp. 478-488.

[3] Steffen, H., "Report on Two Dimensional Stress Strain Behavior of Geomembranes With and Without Friction," Proc Conf. on Geomembranes, IFAI, Denver, CO, pp. 181-185.

[4] Hoekstra, S. E., "Large Scale Bursting Experiments on Geomembranes," RILEM TC 103 Meetings in 1987, 1988 and 1989.

[5] Studiengesellschaft für unterudische Verkehrsanlagen, e.V. (STUVA), "Study to Develop Realistic Tests for Mechanical Behavior of Geomembranes as Base Liner for Landfills," Köln, W. Germany, March, 1988.

[6] van Zanten, R. V., Ed., Geotextiles and Geomembranes in Civil Engineering, A. A. Balkema Publ., Rotterdam, Chapter 12, 1986.

[7] GRI Test Method GM-4, "Three Dimensional Geomembrane Tension Test," Geosynthetic Research Institute, Philadelphia, PA, 1989.

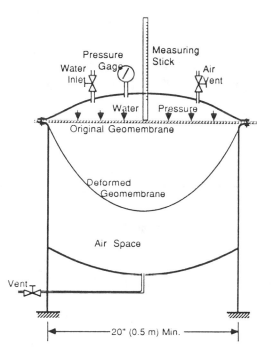

FIG. 1 -- Schematic diagram and photograph of three dimensional axi-symmetric geomembrane tension test apparatus used in the study.

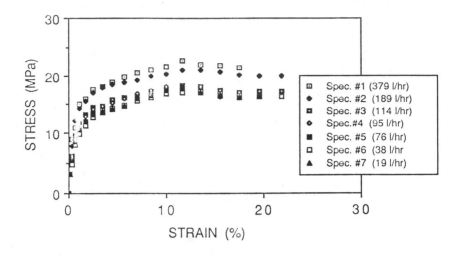

(a) Resulting stress-vs-strain response curves.

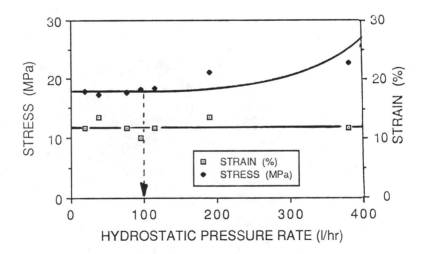

(b) Yield stress and yield strain variation with pressure rate.

FIG. 2 -- Influence of hydrostatic pressure rate on 1 mm thick HDPE geomembranes.

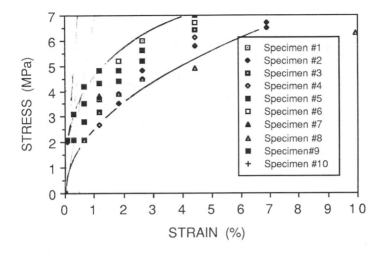

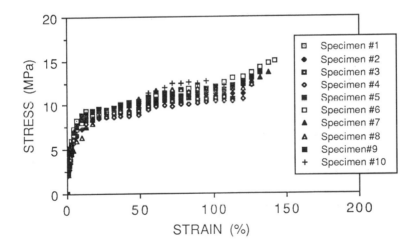

FIG. 3 -- Stress-vs-strain response curves of ten VLDPE geomembrane
          samples taken from the same roll. Upper figure is initial
          portion of curves and lower figure is entire response
          curves.

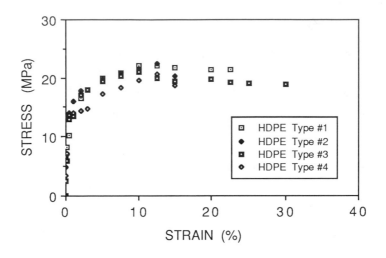

FIG. 4 -- Stress-vs-strain response curves of HDPE geomembranes from various manufacturers.

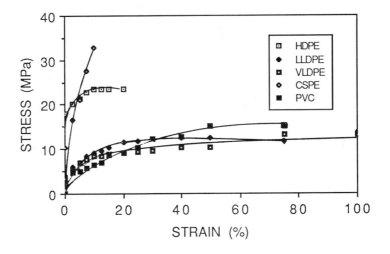

FIG. 5 -- Stress-vs-strain response curves of various types of geomembranes.

APPENDIX*

Strain Calculations

(a) For δ < L/2, assume the geomembrane test specimen to be deformed
     into the segment of a sphere as shown below where "L/2" is the
     radius of the pressure vessel and "δ" is the centerpoint
     deflection.

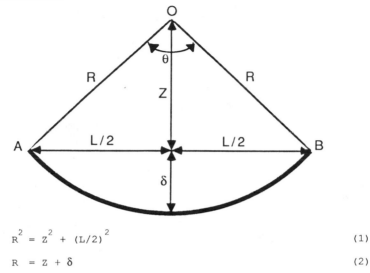

$$R^2 = Z^2 + (L/2)^2 \tag{1}$$

$$R = Z + \delta \tag{2}$$

By squaring equation (2) and substituting it into equation (1)

$$Z = \frac{(L/2)^2 - \delta^2}{2\delta}$$

$$Z = \frac{L^2 - 4\delta^2}{8\delta} \tag{3}$$

now

$$R = Z + \delta = \frac{L^2 - 4\delta^2}{8\delta} + \delta$$

$$R = \frac{L^2 + 4\delta^2}{8\delta} \tag{4}$$

Working with the central angle "θ" and equation (3)

$$\tan(\theta/2) = \frac{L/2}{Z} = \left(\frac{L}{2}\right)\left(\frac{8\delta}{L^2 - 4\delta^2}\right) = \frac{4L\delta}{L^2 - 4\delta^2}$$

$$\theta = 2 \tan^{-1} \frac{4(L)\delta}{L^2 - 4\delta^2} \tag{5}$$

---

* Geometric formulas are from Beyer, W. H., "Standard Mathematical
  Tables," 28th Edition, CRC Press, Inc., Baca Raton, Florida, 1987,
  see Ref 7.

Also

$$AB = R\,\theta \text{ (with } \theta \text{ in radians)} \tag{6}$$

$$AB = \frac{\theta}{360} \cdot 2\pi R = \frac{\theta}{180}\,\pi R \text{ (with } \theta \text{ in degrees)}$$

$$\varepsilon(\%) = \frac{AB - L}{L}(100) \tag{7}$$

Thus, the strain calculations proceed as follows:

$$\bullet\; R = \frac{L^2 + 4\delta^2}{8\delta} \tag{4}$$

$$\bullet\; \theta = 2\,\tan^{-1}\frac{4(L)\delta}{L^2 - 4\delta^2} \text{ (with } \theta \text{ in radians)} \tag{5}$$

$$\bullet\; AB = R \cdot \theta \text{ (with } \theta \text{ in radians)} \tag{6}$$

$$\bullet\; \varepsilon(\%) = \frac{AB - L}{L}(100), \text{ the desired value of strain} \tag{7}$$

Note, that when $\delta = 0$, $R = \infty$, $\theta = 0°$ and $AB = L$, which is to be expected.

(b) For $\delta \geq L/2$, assume the geomembrane test specimen to be deformed in an elliptic shape as shown below.

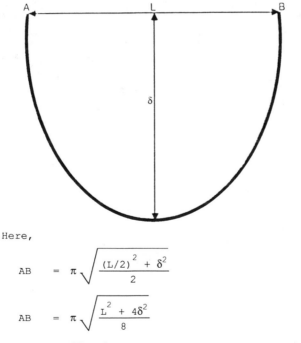

Here,

$$AB = \pi\sqrt{\frac{(L/2)^2 + \delta^2}{2}}$$

$$AB = \pi\sqrt{\frac{L^2 + 4\delta^2}{8}} \tag{8}$$

$$\varepsilon(\%) = \frac{AB - L}{L}(100), \text{ the desired value of strain} \tag{9}$$

Stress Calculations

(a) For $\delta < L/2$, the surface area "S" is assumed to be a portion of a sphere

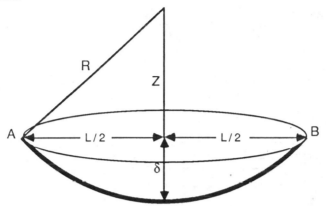

$$S = 2\pi \, R \cdot \delta$$

$$= 2\pi \, \frac{L^2 + 4\delta^2}{8\delta} \, (\delta)$$

$$S = \frac{\pi}{4} \, (L^2 + 4\delta^2) \tag{10}$$

Taking force summation in the vertical direction;

$$S \, p = C \, \sigma' \, t \tag{11}$$

where

    S   = surface area of geomembrane
    p   = applied hydrostatic pressure
    C   = circumference
    σ' = vertical component of geomembrane stress
    t   = geomembrane thickness

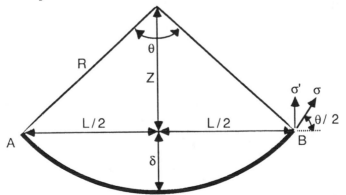

which yields

$$\frac{\pi}{4}\left(L^2 + 4\delta^2\right) p = \pi L \; (\sigma')(t)$$

$$\sigma' = \frac{p(L^2 + 4\delta^2)}{4(L)t}$$

$$\sigma' = \sigma \sin (\theta/2)$$

$$\therefore \; \sigma = \frac{(L^2 + 4\delta^2)p}{4(L) \; t \; \sin (\theta/2)} \tag{12}$$

(b) For $\delta \geq L/2$, assume $\sigma' = \sigma$, thus

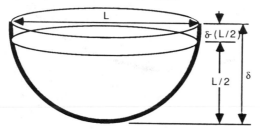

$$\left[\pi L\big(\delta - (L/2)\big) + 2\pi(L/2)^2\right] p = \pi L (\sigma) t$$

$$\sigma = \frac{\left[\pi L (\delta - L/2) + 2\pi(L^2/4)\right] p}{\pi (L) t}$$

$$= \frac{\left[\delta - (L/2) + (L/2)\right] p}{t}$$

$$= \frac{\delta p}{t} \; , \; \text{the desired value of stress} \tag{13}$$

G. Pühringer

PYRAMID PUNCTURE TEST FOR EVALUATING THE PROTECTION
FUNCTION OF GEOTEXTILES

REFERENCE: Puhringer, G., "Pyramid Puncture Test for
Evaluating the Protection Function of Geotextiles,"
Geosynthetic Testing for Waste Containment Applications,
ASTM STP 1081, Robert M. Koerner, Ed., American Society
for Testing and Materials, Philadelphia, 1990.

ABSTRACT: At present, there are inadequate
recommendations to guide the design engineer in the
selection of protection geotextiles in association
with geomembranes. Current recommendations generally
rely on elaborate field experiments or costly
laboratory studies. A relieable engineering
assessment method that will yield useful puncture
values is not currently available. With the pyramid
puncture test, however, testing is possible on
geomembranes and protective nonwoven geotextiles to
determine the magnitude of the protection provided
by geotextiles.

The test apparatus consists mainly of a pyramid piston, an underlaying media and a test press. The
test specimens can be tested unclamped or clamped
between rings. An electric circuit is applied
between the piston and the underlying media to
indicate the puncture force at the point of
perforation.

KEYWORDS: Puncture resistance, electrical circuit,
pyramid piston, geomembrane protection.

1. PURPOSE AND AREA OF APPLICATION

Under these test conditions the puncture test provides a method of testing which permits the evaluation of
the stability requirements necessary in the construction
of waste disposal sites and impoundment (waterproof layer

Pühringer G, is a Technical Consultant of Polyfelt
Ges.m.b.H., St. Peter Str. 25, P.O.B. 675, A-4021 Linz
Austria

and geotextile). This test is intended to be carried out in geotechnical and geosynthetic laboratories. Modified equipment is used which is also suitable for determining the California Bearing Ratio (CBR test). The additional equipment consists of a special piston, adequate fixing clamps and electrical equipment for establishing the puncture load in N. The equipment must allow for a stroke of 100 mm. The puncture test determines the resultant puncture resistance and the subsequent specimen deformation.

2. DEFINITION OF TERMS

The indentation load is the load with which the piston is pressed against the test sample with a constant speed.

The puncture load is that load which causes a perforation of the geomembrane.

The deformation is the longitudinal change in length of the test sample between the inner edge of the fixing ring clamps and the apex of the piston. The deformation is only evaluated with water as the underlying medium and is used only for comparison purposes.

3. EQUIPMENT

3.1 Equipment for preparation of the test

3.1.1 Marking and cutting templates, punch, stamping machine, shears:

The marking and cutting templates can be made of plastic or metal. It is advisable that the screw and guide bolt holes of the test sample be punched out with an 11 mm punch. The test sample itself is cut with a punch or cut out with a large pair of shears.

3.2 The test equipment (Fig 1)

3.2.1 Test set up

A compression press with a reading force accuracy of at least 2 N is necessary. The press must maintain a constant test speed and should be provided with an automatic chart recorder for load vs. deformation. A clamping device for the test sample, a special piston and electrical signal equipment for determining the puncture load are the additional pieces of test equipment needed.

A tension testing machine as is used for the wide stripe tensile test on geotextiles as per DIN 53857/2 or ASTM D4595 can be used. It can be used in the compression mode when the additional equipment described below is used (clamping device, piston and electrical equipment for determining the puncture load).

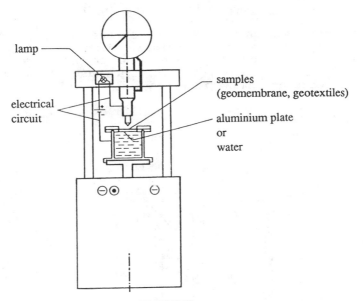

Fig.1: TEST EQUIPMENT

3.2.2 Clamping device

The fixing ring clamps (upper and lower) are described in Fig. 2. The lower fixing ring clamp is provided with a recess on its lower side whose diameter corresponds to the external diameter of the compression base so that it may be fixed in position on the compression base in the compression press. Concentrically arranged grooves are located on the upper side of the lower fixing ring and the lower side of the upper fixing ring for non-slip clamping of the test specimen between the two fixing rings.

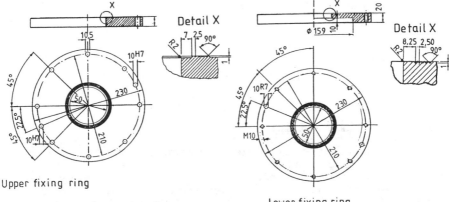

Upper fixing ring

Lower fixing ring

Fig 2: Upper and lower fixing ring
(only for underlaying medium water)

3.2.3 Compression base (Fig. 3a and 3b)

The compression base with which test presses are normally equipped have an internal diamter of at least 150 mm and must be deep enough so that the piston can plunge at least 100 mm into it. The base should be made of rust resistant high grade steel.

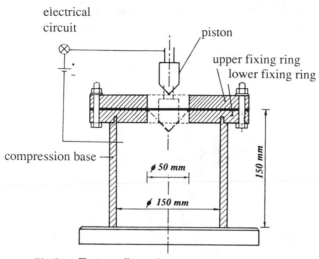

Fig.3a: Test configuration
(for underlying medium water)

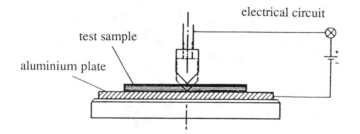

Test configuration
Fig.3b: (for underlying medium aluminiumplate)

3.2.4 Loading piston (Fig. 4)

The loading piston is a cylinder with a pyramid formed apex. The apex is a 4 sided pyramid with an apex angle of 90°. The cylinder (or the base of the pyramid) has a diameter of 25 mm. The apex of the pyramid is rounded off with a radius of R = 0,5 mm, the edges of the pyramid are rounded with a radius of 0,1 mm. The

transitional edge from the base of the pyramid to the
cylinder has a radius of R = 3,0 mm.

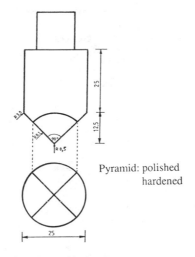

Pyramid: polished
hardened

Fig.4: Pyramid piston

3.2.5 Underlying test media

Either water or an aluminium plate can be used as the
underlying medium for the waterproof layer and geotextile. The water simulates the non-rigid underlying media
found in nature and the aluminium simulates the hard, rigid media. The aluminium plate must be placed on the compression base so that the sealing system (waterproof
layer and geotextile) lies flat on it. In addition the
aluminium plate is to be designed (possibly strengthened
with a steel plate) so that no bending occurs (Fig. 3 b).

3.2.6 Electrical equipment for the determination of the puncture load

An electrical circuit is to be applied on between the
piston and the base medium (water or aluminium plate) for
the exact detection of the puncture load at failure. The
electrical circuit which is closed at the moment when the
waterproof layer is punctured can be indicated by a lamp
signal and thus the effective puncture load can be
recorded.

4. TEST CONDITIONS

With the underlying medium water the advance speed is
v = 50 mm/min.
With the aluminium plate as an underlying medium
V = 1,0 mm/min should not be exceded.

4.1 Geotextiles:

In the case of "base medium water", the edges of the pyramid must be parallel to the machine and cross machine direction of the geotextiles.

5. EVALUATION

The average puncture resistance load in N is to be determined by at least 10 individual tests, both with the underlying medium water as well as with the aluminium plate.

The elongation V is to be given in % and only evaluated with the underlying medium water. Evaluation of elongation is shown in Fig. 5.

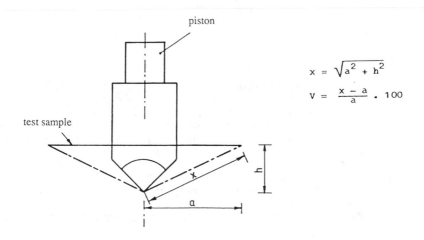

$$x = \sqrt{a^2 + h^2}$$

$$V = \frac{x - a}{a} \cdot 100$$

Fig.5 : calculated deformation
(for underlying medium water)

Where
h=piston movement in mm
a=the distance between the inner edge of the fixing ring and the apex of the pyramid
   piston before deformation of the test sample in mm.
x=the distance between the inner edge of the fixing ring and the apex of the pyramid
   piston at the moment of reaching perforation in mm.
v=elongation in %

6. TYPICAL TEST RESULTS

Puncture tests were conducted on HDPE geomembranes (Table 1) and on needled nonwovens (Table 2)

Tests	HDPE GEOMEMBRANE				
Thickness [mm]	0,50	1,00	1,50	2,00	2,50
Surface	smooth	smooth	smooth	smooth	smooth
Pyramid Puncture					
Strength [kN]	0,20	0,40	0,70	1,05	1,55

Table 1: HDPE geomembranes

Tests	TYPE OF GEOTEXTILE								
	needle punched cont. filam. fabric.[*]								
raw material	pp[1)]	pp	pp	pp	pp	pp	pp	pp	pp
mass DIN 54854 [g/m²]	200	280	400	600	800	1000	1200	1600	2000
CBR-puncture [x-s] DIN 54307 [N]	1900	2750	3400	3200	4100	5500	6000	7800	9000
Strip Elongation [%] (DIN 53857/2)	50/80	50/80	50/80	80/150	80/150	80/150	80/150	80/150	80/150

* Polyfelt TS   1) polypropylene

Table 2: Needle punched fabrics

Fig. 6 and Tab. 3 illustrate the pyramid puncture resistance for an unprotected geomembrane and on HDPE geomembranes protected by needled nonwovens. Geomembranes of various thickness were tested in both cases. The vulnerability of unprotected geomembranes to puncture loads is evident from this figure. For example a 2,00 mm HDPE geomembrane with an 800 g/m² continuous filament nonwoven protection layer has 3 times the puncture resistance of a 2,00 mm thick unprotected HDPE geomembrane.

HDPE Membrane	without nonwoven	Pyramid force of puncture (short-term) (kN)								
		Needled PP continuous filament nonwoven *)								
		200g	280g	400g	600g	800g	1000g	1200g	1600g	2000g
0,50	0,20	0,35	0,50	0,65	0,85	1,10	1,50	1,90	2,80	3,65
1,00	0,40	0,70	0,85	1,10	1,45	1,85	2,40	2,95	4,15	5,25
1,50	0,70	1,10	1,25	1,50	1,95	2,45	3,05	3,60	4,80	6,00
2,00	1,05	1,55	1,80	2,10	2,50	3,10	3,60	4,20	5,45	6,65
2,50	1,55	2,10	2,30	2,65	3,10	3,70	4,55	5,45	7,20	8,90

*)Polyfelt TS

Table 3: Results from the short-term pyramid pressure test conducted on geomembranes
and needled protection nonwovens (base medium: aluminium plate)

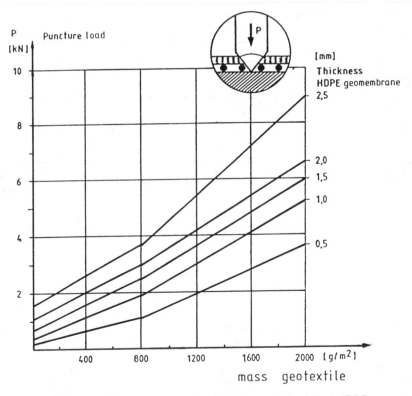

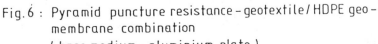

Fig.6 : Pyramid puncture resistance – geotextile / HDPE geomembrane combination
(base medium : aluminium plate)

7 CONCLUSIONS

This test method confirms the concept that geotextiles can be used to enhance the technical performance of geomembranes.

The above results show that the **structure** of the nonwoven (strength, fibre bonding), the **thickness** under load and the **mass** per unit area all influence the "protection function".

The pyramid test method provides a simple index test that allows the protection function of geotextiles to be definable and easily reproducible, that means a direct comparison of different protection geotextiles is possible.

REFERENCES

(1)   Pühringer G., Schneider H., "Analyse der Schutz-funktion von Geotextilien in Kombination mit Dichtungsbahnen", 27th International Conference on Synthetic Fibres, Dornbirn, Austria, Sept. 21 to 23, 1988.

(2)   Pühringer G., "Anwendung von Geotextilien im Deponiebau" Technical Academy of Esslingen, 1987.

(3)   Pühringer G., "The use of geotextiles in the con-structon of landfills", 5. Nürnberger Waste disposal Seminar, April 20 to 21 1989

(4)   Saathoff F., "Kunststoffdichtungsbahnen mit und ohne Schutzvliese unter Punktlasten". 5th professional conference: The safe Landfill, SKZ-Würzburg, February 23, 24, 1989

(5)   Brandl H., "Drainage Systems for the Base Lining of Waste Deposits", Geotechnik 1989/2 page 91.

(6)   Giroud, J.P., "Design of Geotextiles Associated with Geomembranes", Proceedings - Second International Conference on Geotextiles, Vol 1, Las Vegas, 1982, pp. 37-42.

(7)   Koerner, R. et al., "Puncture and Impact Resis-
       tance of Geosynthetics", Proceedings - Third Inter-
       national Conference on Geotextiles", Vienna 1986

(8)   Collins, T.G., Newkirk D.D., "The use of Geotex-
       tile Fabrics in Pond Construction beneath an
       Impermeable Membrane (Geomembrane)", Proceedings
       - Second International Conference on Geotextiles,
       Las Vegas, 1982, pp. 13 - 18.

(9)   Frobel, R.K., Youngblood, W., Vandervoort J.,
       "The composite advantage in the mechanical pro-
       tection of polyethylene geomembranes - a labora-
       tory study", Proceedings Geosynthetics 87 Con-
       ference, Industrial Fabrics Association Inter-
       national, St. Paul, M.N.

Chemical Resistance of Geotextiles and Other Geosynthetics

Karen L. Verschoor, David F. White, and Sam R. Allen

ASSESSMENT OF CURRENT CHEMICAL COMPATIBILITY TEST
METHODS USED TO EVALUATE GEOTEXTILE, GEONETS AND
PIPE.

REFERENCE: Verschoor, K.L., White, D.F. and Allen, S.R., "Assessment
of Current Chemical Compatibility Test Methods Used to Evaluate
Geotextiles, Geonets and Pipe," Symposium on Geosynthetic Testing for
Waste Containment Applications, ASTM STP 1081, Robert M. Koerner,
editor, American Society for Testing and Materials, Philadelphia, 1990.

ABSTRACT: Chemical compatibility testing of geosynthetics material used
in waste management facilities is necessary to assess the compatibility of
these materials with the liquid they will contact. EPA Method 9090 only
addresses specific test methods to be used for geomembranes.
Geosynthetic testing laboratories have adapted this method to test
geotextiles, geonets, and pipe. Current test practices for these materials are
reviewed. Advantages and disadvantages to specific methods are discussed.

KEYWORDS: chemical compatibility, geosynthetic, geotextile, geonet,
pipe, test methods

Introduction and Background

Composite double liner systems with leachate collection and removal
systems are required for hazardous waste facilities and recommended
under minimum technology guidance provided by EPA [1]. Various
geosynthetic materials are typically used in the construction of a waste
management facility [2]. Geomembranes have very low permeability and
often serve as a critical barrier between toxic materials and the
environment. Geotextiles are permeable fabrics that are used for a variety
of functions in waste management engineering. These include: filtration,
drainage, reinforcement, erosion control and separation. Geonets are netlike materials manufactured from extruded polypropylene or polyethylene.
The primary function of a geonet is to provide drainage in the planar
direction. Piping made from polymers are used for collection in leachate
and leak detection systems within a facility. Drainage systems using
geotextiles and geonets are often substituted for natural drainage material
such as stone, gravel or sand. These geosynthetic systems can provide
hydraulic transmissivity equal to the natural materials with significantly less
thickness or weight.

Ms. Verschoor is the Program Manager of the Geosynthetics Testing Program at
TRI/Environmental, Inc., 9063 Bee Caves Road, Austin, Texas 78733. Mr. White and Mr.
Allen are Engineer Scientists at TRI\Environmental, Inc.

To minimize the risk of failure due to geosynthetic material degradation from interaction with liquids they come in contact with, the EPA and state regulatory agencies require data on permit applications that demonstrate the chemical compatibility of material used in a waste management facility. EPA Method 9090 [3] was developed in the early 1980's to give suggested procedures for testing the chemical compatibility of geomembranes with waste liquids. This method recommends that rectangular sheets of geomembrane be immersed in a chemical environment representative of the waste from the proposed facility for a minimum of 120 days at room temperature (22°C) and at an elevated temperature (50°C). Samples are immersed in exposure tanks designed to control loss of volatile components from the exposure media and provide adequate stirring of the media. Various physical properties of the geomembrane are monitored every 30 days. These properties are measured before and after immersion and the results are compared to evaluate the compatibility of the geomembrane with the representative exposure media.

EPA Method 9090 details test methods and exposure conditions to be used when testing the compatibility of geomembrane materials. It does not address test methods to be used when testing any other of the geosynthetic materials. Until very recently, the general practice has been to cite either manufacturers' recommendations or general chemical resistance data from such technical references as plastics applications handbooks to demonstrate the compatibility of the geosynthetics, other than liners, proposed with a specific waste of known composition. These referenced data or recommendations were submitted along with Method 9090 results for the proposed geomembrane liners in the original Part B application.

More recently, the trend has been to perform laboratory testing for all geosynthetics to be used in the proposed containment facility design. Various testing laboratories have adapted EPA Method 9090 to evaluate the chemical compability of geotextiles, geonets and pipe. This paper reviews current bulk physical property test methods used to evaluate these geosynthetic materials to be used in a waste management facility. Analytical techniques used to characterize the base polymer will not be discussed. Limitations to current practices are discussed and specific advantages and disadvantages are given.

Geotextile Testing

Geotextiles vary in the type of polymer used, the type of fiber and the fabric style. Fiber types and fabric styles have been developed for use in a number of general and specific applications. The vast majority of polymers used in geotextile are derived from hydrocarbons. The chemical and environmental endurance required of many of the fabric applications can be traced to the type of polymer used in the geotextile's construction. Fibers used in geotextiles are predominantly made from polypropylene, polyethylene, polyethylene terepthalate and polyamide.

Geotextiles are commonly used in the design and construction of waste management facilities to perform a variety of functions, These include:

- Separation of Layers
- Reinforcement
- Filtration
- Drainage

In geotextile separation, a flexible synthetic barrier is introduced between dissimilar materials such that the integrity and function of both materials can remain intact or be improved. Geotextile reinforcement results for the cooperative effect of the introduction of the geotextile in the strength of a system. Filtration using geotextiles allows for water movement both across the plane of manufacture and through the plane of the fabric itself. Drainage is achieved with geotextiles when the equilibrium system is maintained in the plane of the fabric, allowing free fluid to flow throughout the length of the fabric for an extended time.

The final properties of a specific geotextile product are dependent on the strength and modulus of the individual fibers, how dense the fibers are within the fabric, the interaction of friction between the fibers, and the presences of any coatings or additives between the fibers.

Failure of a geotextile to perform its intended function will result if the material losses enough of its integrity or strength so that it no longer can provide its designed function. A geotextile may lose its ability to filter by the fibers becoming matted or swollen by solvent interaction. Significant clogging may occur from chemical deposits on individual fiber strands.

Testing programs are designed to determine if the geotextile will continue to perform its designed function through the lifetime of a waste facility. Laboratories typically model their testing programs after EPA Method 9090 using the same exposure conditions with significant testing adaptations to account for geotextile characteristics.

Geotextiles demonstrate notable variability within a given roll or product lot in weight, thickness and mechanical properties. This is a normal characteristic and relates to the manufacturing methods used. This variability in properties must be understood and controlled, since chemical compatibility testing depends on the ability to compare properties must be understood and controlled, since chemical compatibility testing depends on the ability to compare properties measured before and after chemical immersion exposure. Interpretation problems could result if, for example, a much thicker section of geotextile were cut into samples and tested at 0 days, while samples tested at a later immersion interval were cut from a thin area.

This problem had been recognized by the commercial laboratories, and four ways of eliminating or controlling variability are in use. The first method involves a pre-screening of cut specimens. First, average roll values for thickness are determined by measuring and weighing 20 specimens cut randomly from various parts of the roll. Then, specimens for each physical property to be measured at each exposure interval are cut from the original roll. The quantity to be cut for each property is increased over needed for the actual testing by a factor of 1.5 or more. After cutting, each individual specimen is weighed. If the weight falls outside one standard deviation from the established roll values, that specimen is rejected and discarded. Only specimens falling within one standard deviation for both weight and thickness are tested. Pre-screened specimens are not marked or identified, and weights are not recorded.

The second method involves recording the weight and thickness of each specimen to be included in the immersion testing. Each specimen to be included in the immersion testing. Each specimen is assigned a coded identification number and individual specimen weights and thickness are recorded into a database which will be kept for reference throughout the project. A numbered tag is attached to each specimen. Specimens are not screened; each weighed specimen is included in the test matrix regardless of weight or thickness. Prior to assigning specimens to exposure baths or intervals, all cut, weighed and tagged samples are shuffled physically so that weights and thicknesses throughout the range are evenly distributed. When test data are reported, original weights and thickness for the unexposed specimen database are included with the test results. Outliers may be identified and considered in the context of any apparent trends in the data.

The third method involves cutting all specimens required for baseline and immersion interval testing from one selected region across the supply roll of geotextile. Weight and thickness compliance with minimum roll value specifications is verified before testing begins. This method recognizes that due to the manufacturing procedures used, product weight and thickness tend to vary in a consistent way across any given roll.

The fourth method consists of testing seven replicates for each mechanical property test, and rejecting two of the seven as outliers so that the reported test values is an average of five replicates. No pre-screening or sample selection is done prior to testing.

There is some controversy over whether geotextile specimens should be tested in wet or dry condition. To completely dry the specimen would require an extended period of time, or would mean use of a drying oven which could introduce exposures outside the realm of the chemical effects that are to be investigated. A further problem arises in deciding how to determine when a specimen is truly "dry." Exposed specimens may be rinsed with water after removal from the exposure tanks to remove gross contamination, or the specimens may be tested while still coated with leachate after superficial drying. It has been argued that the second approach better simulates service conditions, but it introduces severe problems with equipment contamination and safety.

A general consensus has been reached that the best approach is to test geotextile samples in damp condition after a water rinse to remove gross contamination. The practice being followed is to measure baseline, or unexposed properties in damp condition as well, so that a proper comparison may be made. Unexposed samples are rinsed in deionized water and patted dry prior to testing. This procedure is now believed to be in use at each of the commercial testing laboratories.

The following tests have been employed by various manufacturers and testing labs to determine changes in a geotextile's capacity to perform it intended function:

> Mass per Unit Area
> Thickness
> Tensile Properties:
> - Grab Strength
> - Strip Tensile
> - Wide Width Tensile Strength
> Puncture
> Trapezoidal Tear
> Hydrostatic Burst Strength
> Water Permeability (Permittivity)
> Hydraulic Transmissivity
> Apparent Opening Size

Mass per unit area is the proper term for weight of a geotextile. Sometimes referred to as "basis weight", it is usually given in units of ounces per square yard (oz/yd^2) or grams per square meter (g/m^2). The mass of the fabric should be measured to the nearest 0.01% of the total sample mass and the length and width should be measured under zero fabric tension.

Monitoring weight change as a function of exposure, as required for liners in Method 9090, is difficult for geotextiles since it would require that each geotextile specimen be fully dried prior to weighing and subsequent destructive testing. Some laboratories have attempted to resolve this problem by including specimens specifically assigned for weight and dimensional determinations which are not destructively tested.

The thickness of a geotextile is measures as the distance between the upper and lower surfaces of the material measured as a specified pressure. Thickness of commonly used geotextile range form 10 to 300 mils.

Tensile properties are analyzed by grab, strip, or wide width tensile tests. Grab tensile tests utilize a small sample size gripped by small 1" x 3" faces making them susceptible to side wall effects. To avoid this problem some labs prefer a modified strip tensile test, also used for geotextile seams, using a small sample size gripped along the full width of the specimen. Wide width tensile tests use 8" wide coupons gripped along the full specimen width. This test is commonly used to determine seam strength. The large sample size required for the wide width tensile test prohibit testing with multiple specimens. In all tensile tests, a continually increasing load is applied longitudinally to the sample and the test is continued to rupture.

The puncture test assesses geotextile resistance to objects such as sticks and rocks under quasi-static conditions. A blunt-ended metal rod (5/16 in. dia) is pushed through the fabric which is firmly clamped in a empty cylinder (1 3/4 in. dia.) by a compression testing apparatus. The resistance to puncture is then measured in pounds force.

The trapezoidal tearing load is the force requires to successfully break individual fibers in a fabric. The fabric is inserted into a tensile testing machine on the bias so that the fibers tear progressively. An initial 5/8 inch cut is made to start the process, The individual fibers are actually stressed rather than the entire fabric system.

Hydrostatic burst resistance testing loads fabric out of plane and stresses them to failure. The fabric is deformed in various ways by use of rubber diaphragm until the central portion of the geotextiles, lying along the minor axis of the fabric, yields to strain. Alternately, water under hydrostatic pressure may be used to burst the specimen.

Water permeability or permittivity testing quantities water it passes through a 4" diameter geotextile specimen in an isolates condition. In the "falling head" permittivity test, a column of water is allowed to flow through the geotextile and reading of head changes versus time are taken. The "constant head" permittivity test maintains a minimum head of two inches on the geotextile throughout the test. This test is frequently performed for geotextile chemical compatibility studies because of its small sample size requirements and the relative consistency in test results. The transmissivity or in-plane permeability of the geotextile quantifies the planar drainage performance of the fabric. As with permittivity, the variation of the fabric thickness under compressive load must be considered.

The apparent opening size of a geotextile is determined by mounting a specimen in a sieve frame with glass beads placed on the geotextile surface. The geotextile and frame are shaken laterally so that the jarring motion will induce the beads to pass through the test specimen. The procedure is repeated with various size beads until the sample's apparent opening size is determined. This test is unattractive as a chemical compatibility index test because it must be performed with dry samples. Labs have reported difficulties with variability in test results caused by retained solids from leachate and static electricity.

Individual geotextile fiber tensile strength has been measured by applying any of a number of standard methods used in the textile industry. Fiber testing has some very significant advantages for chemical compatibility testing. It requires that only very small bulk of geotextile be exposed to a chemical or leachate medium. Fiber testing may provide better reproducibility and precision for some products as compared with other index tests, since the variability caused by manufacturing processes is eliminated.

Geonet Testing

Geonets, unlike geotextiles, are relatively stiff, netlike materials with large open spaces (0.9 to 5.0 cm) between the structural ribs. Geonets serve primarily in a drainage role although they add some structural stability to any application in which they are used. Geonets are generally extruded and have three dimensional structures. Non deformed nets are used primarily as a core material to provide planar flow in drainage systems. Geonets are primarily constructed of polyethylene and polypropylene as are geomembranes.

Failure of a geonet in a leachate collection or leak detection system can caused by crushing of the geonet resulting in either a total shutdown of leachate flow or a failure to meet the hydraulic transmissivity or detection time requirements. Decreases in flow time rate with time may results for intrusion of an overlying geotextile into the open area od a geonet, compression of the geonet, build up of air within the test specimen, or a combination of these and other factors. It is often difficult to isolate a single component for performance evaluation and relate this to chemical compatibility.

The most common method used to evaluate the performance of a geonet is the hydraulic transmissivity test (ASTM D4716). This test allows the user to monitor changes in hydraulic conductivity in the planar direction over time under a constant normal compressive stress. Performance of the specimen is determined by measuring flow rate.

Other tests for geonets include:

> Mass per Unit Area
> Aperture Size
> Volatiles and Extractables Content
> Tensile Properties:
> Wide Width Tensile
> > Strip or narrow width tensile
> Single rib tensile
> Junction node strength
> Evaluation of Compression
> Creep Compliance

Physical properties of geonets are characterized by measuring dimensions and aperture of samples. Aperture of a geonet or geogrid is determined by measuring the length in each principal direction of a single opening and is critical to the characteristic flow properties of geonets.

Strength properties of geonets and geogrids are typically determined by testing the tensile properties of a product. Wide width tensile tests are typically avoided as part of chemical compatibility studies due to large sample size requirements. Although no formal test method exist, the majority of laboratories utilize a strip tensile test employing a small test specimen that is gripped across the entire width of the coupon. In general, only one of the four possible directions is tested. The strip tensile test is preferred to standard methods such as wide width tensile strength because of sample size. One laboratory uses a quality control specification developed by a geonet manufacturer for strip tensile testing of net in one direction. No ASTM standard exists for narrow strip tensile testing of geonet.

Volatiles and extractable content tests are used to measure the amount of volatiles and extractables in a material before and after exposure to leachate. The procedures are commonly performed for geomembrane liners as part of Method 9090 testing. Samples are weighed, heated and weighed again to determine percent weight loss. any change in weight during the heating cycle is due to loss of volatile substance(s) which were absorbed form the leachate. If a weight loss is observed by virtue of exposure to leachate and extraction of a soluble component is indicated. Labs report problems with this test due to introduced error, since quantities being measures are considerably smaller than the inherent error introduced during testing.

Compression and creep compliance test are commonly used to evaluate geonets. Although there are no generally accepted standard methods Geosynthetics Research Institute has developed methods that the industry is using until standards can be established.

Pipe Testing

Plastic pipes, often slotted or perforated, are used for leachate collection and removal in primary and secondary systems. Failure of pipes can occur if the compressive strength of the pipe is reduced due to chemical degradation so that he pipe is crushed under it's load and leachate flow is inhibited.

A wide diversity of methods for testing pipe have been evolved and published by ASTM, the Plastic Pipe Institute, the Gas Research Institute and the National Sanitation Foundation. The following tests have been employed by various manufacturers and testing labs to determine changes in a pipe's capacity to perform these functions as a result of chemical exposure:

- Dimensions and weight
- Pipe Stiffness
- Tensile Properties
  - strip tensile
  - ring tensile

- Three Point Bend
- Environmental Stress Crack Resistance
- Hydraulic Burst Strength

Pipe dimensions are determined by measuring specimen length, wall thickness, and inner and outer diameter. Due to the wide variety of solid, corrugated and perforated pipe products currently used in hazardous waste applications, the specific procedures and conventions used for physical property testing have varied. Corrugated pipes often vary greatly in wall thickness. Sample preparation in sometimes difficult when attempting uniformity in sample length. Typically, laboratories will etch identification marks in pipe test coupons so that dimension data may be generated at the same location for each coupon during testing.

Strength properties of pipe are typically characterized by stiffness or compression resistance testing since this is recognized as the most critical performance requirement. A sample is static loaded between two parallel plates until a 30 percent deformation is observed. Stress at various deformations is recorded. There has been some debate as to the appropriate sample size needed for this test. While small 3" long pipe samples are easier to accommodate during chemical exposures, they may exhibit end effects during loading. Manufacturers have reported that such variables as the cut end profile and quality of cut have a large influence on results for samples shorter than required in the standard. Larger samples reduce variability introduced by reduced sample dimensions but require additional exposure room in tanks and reduce the number of test replicates which can be tested.

Some manufacturers have considered developing a strip bending test to monitor pipe strength characteristics. One possible test is as follows: An arc shaped test coupon is cut from a sample pipe to be of consistent length and width. The test coupon is allowed to rest on two supports and is loaded by means of a loading nose midway between the supports. As in a full size compression test, load versus deflection curves are generated. Interviewed participants agreed that this test would be attractive to chemical compatibility studies because of decreased exposure requirements and increased replicate testing. The three-point bend loading more closely simulates the pipe stiffness test than do options such as tensile or creep compression tests. It is important to note that, in waste containment applications, pipes will be loaded in compression only.

Tensile properties of pipe materials are often characterized by extracting a strip tensile test coupon from a pipe wall. Sample preparation may be difficult with perforated or corrugated pipes but procedures can usually be developed for specific pipe dimensions. Ring tensile tests have also been performed, but they apply tensile stresses which do not characterize the typical waste containment applications: leachate recovery, collection and removal.

Environmental stress crack resistance (ESCR) of pipes is determined by extracting a strip coupon form the pipe wall, notching it with a specified cut, and mounting it in a metal rack. Samples are usually exposed to Igepal CO-630, a common surfactant, at elevated temperature. Samples are observed for propagation of the cut at various times. Thick wall pipes make sample preparation difficult but extraction methods may be modified and special stressing apparatus built to accommodate this analysis. The test is mainly applicable to polyethylene pipe, and is widely performed as part of Method 9090 testing for polyethylene geomembrane liners.

Discussion

The tests described in the previous section represent those most often used to characterize geosynthetics other than geomembranes. Table 1. outlines these test methods, their applicable standards, and advantages and disadvantages of their use in chemical compatibility studies.

Since standardization has been lacking for chemical compatibility testing of geotextile, geonet and pipe, the selection of test methods for evaluation of property changes caused by leachate exposure has been left up to the individual facility owner or designer. From an engineering standpoint it would be desirable to learn whether design or functional performance properties of the geosynthetic component would be affected by extended leachate exposure. However, it is not practical in most cases to measure performance properties, and it has become standard practice to select index tests for inclusion in chemical compatibility test programs. The goal in selecting and evaluating results of index tests is to determine whether the leachate is interacting with the geosynthetic product in a way that might degrade the material or its ability to perform an intended design function over the long term.

Criteria used to select index or performance tests for inclusion in chemical compatibility studies can be summarized as follows:

- The test must provide reproducible results and be relatively independent of experimental factors or operator-introduced errors.

- The test must require relatively small individual specimens so that leachate volume can be minimized, and allow a sufficiently large number of replicates (especially critical for geotextiles).

- The test should not measure a property or simulate an exposure that would be completely uncharacteristic of the proposed design function.

Although there is general agreement on the criteria to be used to select appropriate test methods, the specific methods used vary between the various laboratories. It is recommended that the industry work to develop standards within the framework of consensus standards orginazation such as ASTM, and that regulators cite these industry standards when they become available.

ACKNOWLEDGEMENT

TRI/Environmental, Inc. wishes to acknowledge the support of the Environmental Protection Agency through its subcontract with Southwest Texas State University.

REFERENCES

[1]   "Minimum Technology Guidance on Double Liner Systems for Landfills and Surface Impoundments, Design and Construction and Operation", USEPA Report No. EPA/530-SW-85014, U.S. Environmental Protection Agency, Washington D.C.

[2]   Koerner, R., "Designing with Geosynthetics", Drexel University. Philadelphia, Pennsylvania

[3]   EPA Method 9090, "Compatibility Tests for Waste and Membrane Liners", in EPA SW-846, Test Methods for Evaluating Solid Waste, U.S. Environmental Protection Agency, Washington, D.C.

TABLE 1
PHYSICAL PROPERTY TEST METHODS FOR CHEMICAL COMPATIBILITY

Description	Standard	Sample Configuration	Advantages for Chemical Compatibility	Disadvantages for Chemical Compatibility
Grab Strength	ASTM D4632	4"x8" rectangle (geotextile)	Small sample size. Universal test	Gripping procedure causes variability in test results caused by edge effects.
Strip Tensile	ASTM D751	2 inch wide strip (geotextile;geonet)	Small sample size. Gripping full width of sample decreases variability caused by side wall effects. Good reproducibility reported.	Currently no standard specifically written for geosynthetics.
Wide Width Tensile	ASTM D4595	8 inch wide strip (geotextile;geonet)	Large size decreases variability caused by dimensional fluctuations along roll width.	Large sample size decreases number of test replicates. Exposure requires large volume of leachate.
Fiber/Yarn Tensile	ASTM D3822	Single filament or yarn (geotextile)	Small sample size. High precision. Universal test among geotextile manufacturers.	Many replicates requires. Untried in chemical compatibility.
Puncture	ASTM D4833	4" diameter circle (geotextile)	Small sample size. Universal test.	High variability in test results caused by dimensional variations inherent in product.
Trapezoidal Tear	ASTM D4533	3"x8" notched rectangle (geotextile;geonet)	Small sample size. Universal test.	High variability in test results caused by dimensional variations along roll width and sample prep inconsistency.
Burst Resistance (Mullen)	ASTM D3786	4" diameter circle (geotextile)	Small sample size. Repeatable, reliable test.	

TABLE 1

PHYSICAL PROPERTY TEST METHODS FOR CHEMICAL COMPATIBILITY (cont'd)

Description	Standard	Sample Configuration	Advantages for Chemical Compatibility	Disadvantages for Chemical Compatibility
Permittivity	ASTM D4491	4" diameter circle (geotextile)	Small sample size. Universal test.	Some inter-lab variability reported due to apparatus design.
Hydraulic Transmissivity	ASTM D4716	12"x12" square (geotextile)	Large sample size decreases effects of dimensional variability. Models design function of product.	Large sample size requires large volume of leachate for exposure. Some inter-lab variability reported due to apparatus design.
Apparent Opening Size	ASTM D4751	12" diameter circle (geotextile)	————	Sample must be tested dry. Variability in test results caused by static electricity and retention of leachate.
Melt Flow Index	ASTM D1238	3-6 grams (geonet)	Small sample size. Universal test	Necessary melt temperature may cause decomposition of some polymers. Results may not correlate to service properties.
Solution Viscosity	————	1-3 grams (any geosynthetic)	Small sample size.	No formal standard associates with geosynthetics community.
Aperture Size	————	Single opening (geonet)	Small sample size. Common specification in geonet industry.	No formal standard
Volatiles and Extractables	EPA 9090	(any geosynthetic)	Small sample size. among geotextile manufacturers.	High variability in test results due to large inherent error in test method.

TABLE 1
PHYSICAL PROPERTY TEST METHODS FOR CHEMICAL COMPATIBILITY (cont'd)

Description	Standard	Sample Configuration	Advantages for Chemical Compatibility	Disadvantages for Chemical Compatibility
Single Rib Tensile	———	Single rib (geonet;geogrid)	Small sample size enabling several test replicates.	No formal standard
Junction Node Strength	———	Single node (geonet,geogrid) (geotextile;geonet)	Small sample size enabling several test replicates.	No formal standard
1" Ball Burst	ASTM D751 (modified)	4" diameter circle (geonet)	Small sample size.	High variability in test results.
Creep Compliance	———	Square sample (geonet)	Generates results for application to design. Small sample size Accurate product thickness data provided.	No formal standard
Pipe Stiffness	ASTM D2412	3-12" long specimens (pipe)	Universal test. Small sample size with 3" specimens. Good reproducibility claimed with >12" specimens.	Test results vary for short specimens due to side wall effects. Pipe geometrics require large volumes of leachate for exposure.
Strip Tensile	ASTM D638 (modified)	1/2"-1" wide strip or dumbell (pipe)	Small sample size. Universal test.	Samples must be machine cut for thick wall pipe. Modifications are necessary for corrugated and slotted pipe.

TABLE 1
PHYSICAL PROPERTY TEST METHODS FOR CHEMICAL COMPATIBILITY (cont'd)

Description	Standard	Sample Configuration	Advantages for Chemical Compatibility	Disadvantages for Chemical Compatibility
Ring Tensile	ASTM D2290	1"-6" wide tube section (pipe)	Small sample size. Universal test	Test results vary for short specimens due to side wall effects. Pipe geometry requires large volume of exposure liquid. Mode of loading not characteristic of service.
Three Point Bend	———	1"-6" wide arc (pipe/geosynthetic)	Small sample size. Simulates loading modes experienced in installation.	No formal standard
Environmental Stress Cracking	ASTM D1693	1/2"-3" arc (pipe)	Small sample size.	Test specimens must be machine cut from thick wall pipe.
Hydraulic Burst Strength	ASTM D1180	——— (pipe)	———	Test does not monitor stresses simulating field installation and usage.
Carbon Dispersion	ASTM D3015	Slide preparation (pipe)	Test does not monitor relative changes experienced in field applications.	No formal standard

C. Joel Sprague

LEACHATE COMPATIBILITY OF POLYESTER NEEDLEPUNCHED NONWOVEN GEOTEXTILES

REFERENCE: Sprague, C. J., "Leachate Compatibility of Polyester Needlepunched Nonwoven Geotextiles", Geosynthetic Testing for Waste Containment Applications, ASTM STP 1081, Robert M. Koerner, editor, American Society for Testing and Materials, Philadelphia, PA, 1990.

ABSTRACT: The durability of geotextiles -- that is, the ability of the geotextile microstructure to resist degradation -- is an important criterion for geotextile selection. The durability of a geotextile is influenced by its unique characteristics, such as polymer type, additives used, processing history, bonding mechanism, fiber element geometry, and fabric construction. Therefore, product specific testing is the only dependable approach to characterize the durability of a given geotextile. This paper details chemical and leachate compatibility testing of polyester needlepunched nonwoven geotextiles.

KEY WORDS: Leachate, polyethylene terephthalate, hydrolysis, durability, chemical resistance, crystallinity.

INTRODUCTION AND BACKGROUND

The objective of the testing program detailed in this paper was to use traditional physical property testing to monitor aging of geotextile fabrics exposed to deionized water, sodium hydroxide solution (pH-12), and calcium hydroxide solution (pH-12.4) and compare these results to results obtained from samples that had been aged in leachates from three hazardous waste facilities. Calcium hydroxide was included because it is known to degrade polyester through hydrolysis at a faster rate than sodium hydroxide.

Mr. Sprague (formerly with Hoechst Celanese Corporation) is a Geosynthetics Engineer at Nicolon Corporation, 3500 Parkway Lane, Suite 500, Norcross, Georgia 30092.

In an effort to better understand chemical effects on polyester geotextiles, Hoechst Celanese Corporation contracted Texas Research Institute to study both staple and continuous filament geotextile samples aged in a controlled environment conducive to hydrolysis of polyester using physical property testing to characterize the strength and resultant serviceability of the geotextile product.

The term polyester used herein refers to polyethylene terephthalate (PET). Hydrolysis of PET basically is the reverse reaction of the synthesis of PET, i.e., a long chain linear molecule is split again by an H_2O molecule resulting in a scission of an ester bond. One scission/molecule cuts the molecular weight by half [1].

Three geotextile fabrics were studied: 1114, a 0.120 kg/m^2 continuous filament polyester nonwoven fabric; 1155, a 0.550 kg/m^2 continuous filament polyester nonwoven fabric; and 7155, a 0.550 kg/m^2 staple polyester nonwoven fabric. Continuous filament geotextiles are produced from fibers that are drawn from melted polymer extruded through dies or spinnerets. The fibers are continually extruded, drawn, cooled and distributed to form a uniform web. The web is then bonded by a needlepunching process. Staple geotextiles are produced from staple fibers or filaments cut in short lengths from previous fiber processing. These staple fibers commonly range in length from one to eight inches and are formed and cut at a different time and location from the geotextile fabric processing.

The physical property test methods used to monitor geotextile aging were selected based on those typically used for EPA Method 9090 chemical compatibility testing. EPA Method 9090 measures changes in physical properties of geotextile samples after exposure to leachates under specified conditions. In addition to performing grab strength, elongation, and puncture resistance testing, exposed samples were shipped to Hoechst Celanese Corporation for the performance of Mullen burst, permittivity, and solution viscosity testing. The solution viscosity (SV) of the fibers was used as an indicator of the molecular weight of the fiber polymer. All baseline data is shown in Table 1.

METHODS

Sample Preparation

All geotextile specimens were die cut in either 0.1m diameter circles or 0.1m by 0.2m rectangles. All specimens were randomly mixed and subsequently selected for test conditions. All specimens were rinsed with deionized water to prevent contamination of test equipment. After drying, the specimens were allowed to equilibrate at room temperature and ambient humidity for 24 hours before weights and thickness measurements were recorded. All samples were tagged with a tantalum tag on a stainless steel pin.

Exposure Conditions

Exposure of the test samples was performed following the guidelines outlined in EPA Method 9090. Polyethylene tanks used for the exposure were maintained at 22°C ± 2°C. Per EPA Method 9090 procedures, the 50°C tanks were fitted with a reflux condenser to control loss of chemicals through evaporation.

Samples of 1114, 1155 and 7155 were exposed to three different chemicals: deionized water; sodium hydroxide solution (pH-12); and a saturated calcium hydroxide solution (pH-12.4). The alkaline solutions were replaced with freshly prepared solutions weekly.

Samples of 1114 and 7155 were exposed to leachates from three hazardous waste landfills. The three leachates used for these sample exposures were: "L", from a Louisiana facility; "O", from an Ohio facility; and "N", from a New York facility. A qualitative characterization of the leachates is given in the Appendix. The leachate "N" differed from the other leachates primarily in alkalinity. This landfill used lime for fill and the resultant leachate had a pH of approximately 12 when it was delivered to Texas Research Institute. The pH of the other leachates was not determined.

Samples were removed for testing at 120 days for the leachate exposures, at 30, 60, 90 and 120 days for the deionized water exposures, at 1, 2, 4 and 6 weeks for the calcium hydroxide solution exposures, and at 1, 2, 4, 6, 10 and 14 weeks for the sodium hydroxide solution exposures. The more frequent testing for the chemical exposures was planned to more accurately assess the corresponding greater expected degradation.

Physical Property Testing

Heat dried, unexposed and exposed specimens of each of the three geotextile samples were tested for grab strength and puncture resistance following ASTM Standard Methods D4632 and D4833, respectively. Mullen burst and permittivity testing was done at Hoechst Celanese in accordance with ASTM D3786 and D4491, respectively.

RESULTS

Tables 2 and 3 summarize the results from testing 1114 in the various chemicals and leachates. Table 4 summarizes the results from testing 1155 in the various chemicals. Tables 5 and 6 summarize the results from testing 7155 in the various chemicals and leachates. Average percent changes were calculated relative to baseline values for the tests performed at Texas Research Institute. For the tests performed at Hoechst Celanese Corp., average percent changes were calculated relative to "typical" values given by the manufacturer, not actual baseline data generated from the samples tested.

TABLE 1 - - Baseline Data for Geotextiles Tested

Type	Puncture Strength D4833 (kN)	Grab Tensile Strength D4632 (kN)	Grab Tensile Elongation D4632 (%)	Mullen Burst Strength D3786 (kPa)	Permittivity D4491 $(1/s/m^2)$	SV
1114	0.289	0.667	65	1586	136	840
1155	1.023	2.891	70	6205	51	840
7155	0.756	1.468	95	3792	51	700

TABLE 2 - - Effect of Temperature and Leachate on 1114

	Time (Weeks)	% Change Puncture	% Change Grab Strength	% Change Elongation
H_2O 22°C:	17.2	12	7	-24
H_2O 50°C:	17.2	7	18	-21
"L" 22°C:	17.2	24.6	5.9	-18.4
"L" 50°C:	17.2	17.5	8.1	-19.7
"O" 22°C:	17.2	24.6	8.8	-15.8
"O" 50°C:	17.2	17.5	7.4	-14.5
"N" 22°C:	17.2	15.8	7.4	-23.7
"N" 50°C:	17.2	22.8	3.7	-21.1

TABLE 3 - - Effect of Temperature and Chemical on 1114

Time (Weeks)	% Change Puncture	% Change Grab Strength	% Change Elongation	% Change Mullen Burst	% Change Permittivity	% Change SV
H₂0 22°C:						
4.3	12	4	- 4	5	-18	- 1
8.6	30	1	-24	19	- 1	- 1
12.9	16	6	- 7	4	-13	0
17.2	12	7	-24	6	-11	1
H₂0 50°C:						
4.3	18	- 1	- 9	4	-18	- 1
8.6	18	13	-14	7	-17	- 3
12.9	21	4	- 4	9	- 2	- 2
17.2	7	18	-21	0	- 6	- 2
NaOH 22°C:						
1	23	0	- 7	11	-17	- 3
2	19	10	- 8	9	-16	- 2
4	14	13	-18	7	-11	- 1
6	16	7	-26	0	-16	- 2
10	2	- 3	-20	7	-17	- 1
14	9	5	-24	- 1	-16	- 1
NaOH 50°C:						
1	9	0	- 4	1	-19	- 1
2	14	1	-12	0	1	N/A
4	0	4	-25	2	-15	- 2
6	12	7	-21	8	-14	- 1
10	- 2	5	-25	- 1	6	- 2
14	7	1	-25	4	3	- 2
Ca(OH)₂ 22°C:						
1	14	- 4	-12	- 2	-14	- 3
2	5	2	-24	- 6	-19	- 2
4	-14	-15	-30	N/A	N/A	- 3
6	-23	-25	-29	-17	-20	- 2
Ca(OH)₂ 50°C:						
1	- 2	-10	-14	-13	-14	- 2
2	-35	-40	-28	-31	-17	-21
4	-63	-69	-30	N/A	N/A	-45
6	F/D	F/D	F/D	F/D	F/D	-73

N/A: Data not available
F/D: Fabric destroyed

TABLE 4 - - Effect of Temperature and Chemical on 1155

Time (Weeks)	% Change Puncture	% Change Grab Strength	% Change Elongation	% Change Mullen Burst	% Change Permittivity	% Change SV
H₂0 22°C:						
4.3	7	- 1	9	5	6	- 4
8.6	5	3	- 7	6	- 1	- 4
12.9	9	- 1	2	5	2	- 3
17.2	4	6	-10	5	- 3	- 3
H₂0 50°C:						
4.3	2	5	1	0	3	- 4
8.6	4	1	- 9	5	2	- 5
12.9	3	- 4	5	4	7	- 4
17.2	0	4	-12	1	3	- 4
NaOH 22°C:						
1	- 6	- 9	0	4	10	- 4
2	- 5	2	- 1	5	1	- 4
4	2	- 6	1	6	2	- 3
6	- 8	6	-11	3	- 2	- 5
10	- 7	- 4	- 9	- 3	3	- 4
14	- 3	- 4	- 5	6	3	- 3
NaOH 50°C:						
1	- 1	- 1	- 5	4	- 3	- 5
2	- 4	1	- 7	3	- 2	- 5
4	- 7	5	- 5	1	- 3	- 5
6	0	12	9	4	7	- 5
10	- 5	2	- 9	4	6	4
14	-11	0	- 4	2	- 4	- 5
Ca(OH)₂ 22°C:						
1	- 8	- 3	- 4	- 5	2	- 5
2	-19	- 7	-13	- 3	-15	- 5
4	-27	-21	- 5	N/A	N/A	-12
6	-25	-28	- 9	-24	2	- 6
Ca(OH)₂ 50°C:						
1	-15	-17	-18	-10	-11	- 6
2	-36	-34	-17	-30	-14	-27
4	-60	-64	-17	N/A	N/A	-21
6	-89	-92	-38	-47	6	-65

N/A: Data not available

TABLE 5 - - Effect of Temperature and Chemical on 7155

Time (Weeks)	% Change Puncture	% Change Grab Strength	% Change Elongation	% Change Mullen Burst	% Change Permittivity	% Change SV
H₂0 22°C:						
4.3	1	4	-16	-10	- 1	- 1
8.6	- 3	- 3	-30	- 5	- 8	- 1
12.9	-17	15	-14	- 9	-18	2
17.2	-11	11	-25	- 4	-12	1
H₂0 50°C:						
4.3	- 9	3	-16	- 9	- 5	- 2
8.6	-10	2	-23	- 8	- 9	- 2
12.9	-15	3	-16	- 9	N/A	1
17.2	- 7	4	-28	-15	-10	- 1
NaOH 22°C:						
1	-24	1	-14	-18	4	0
2	-15	- 3	-22	-15	-12	0
4	-22	0	-34	-15	-16	- 1
6	-20	- 4	-33	-23	12	N/A
10	-25	-13	-29	-22	-13	1
14	-18	- 8	-28	-22	8	- 0
NaOH 50°C:						
1	-28	- 9	-27	-22	10	- 1
2	-28	-12	-30	-19	- 3	- 2
4	-17	-10	-35	-24	-15	- 2
6	-24	- 9	-35	N/A	N/A	- 2
10	-31	-12	-29	-26	-16	0
14	-28	-10	-30	-24	1	0
Ca(OH)₂ 22°C:						
1	-20	-15	-29	-23	- 8	N/A
2	-31	-18	-39	-31	- 8	1
4	-45	-26	-37	N/A	N/A	- 2
6	-48	-34	-37	-40	- 7	- 2
Ca(OH)₂ 50°C:						
1	-40	-24	-30	-27	- 4	- 2
2	-53	-39	-37	-48	-16	- 2
4	-65	-59	-31	N/A	N/A	- 9
6	-78	-72	-45	-69	-14	-13

N/A: Data not available

TABLE 6 - - Effect of Temperature and Leachate on 7155

	Time (Weeks)	% Change Puncture	% Change Grab Strength	% Change Elongation
H₂0 22°C:	17.2	-11	11	-25
H₂0 50°C:	17.2	- 7	4	-28
"L" 22°C:	17.2	-8.3	4.7	-11.5
"L" 50°C:	17.2	-13.0	0.0	-12.6
"O" 22°C:	17.2	-16.1	4.7	-14.9
"O" 50°C:	17.2	-29.7	-17.7	-19.5
"N" 22°C:	17.2	-20.8	-0.4	-23.0
"N" 50°C:	17.2	-41.7	-26.7	-24.1

DISCUSSION

Polyethylene terephthalate (PET), commonly called polyester, molecules consist of flexible aliphatic portions and relatively stiff aromatic portions. Crystallization occurs upon drawing of the fiber, creating oriented crystalline molecules consisting of both crystalline and amphorous regions. Aging in an aqueous environment results in an increase in crystallinity in the amorphous regions as a result of initial hydrolysis shortening of some randomly oriented molecules enabling them to crystallize (chemical crystallization) [2]. This increase in crystallinity affects fiber strength and fiber elongation [3]. As hydrolysis continues, it leads to more chain scissions and a corresponding reduction in molecular weight which has a strong effect on fabric strength.

It is known that PET is susceptible to alkaline hydrolysis and that such hydrolysis occurs topically at the amorphous portion of the molecule [1]. Different forms of PET are more resistant to hydrolysis. In general, a higher molecular weight polymer (i.e., higher SV) is more hydrolysis resistant because of its higher degree of crystallinity than a lower molecular weight form of PET [4].

When PET is aged in an environment that has both chemical and thermal stresses, the following things are likely to occur: thermal relaxation of the individual fibers, changes in polymer crystallinity, and hydrolysis of the polymer chain. In this study, it was believed that hydrolysis was the major cause for the decline of physical properties and that the hydrolysis occurred at a faster rate at higher temperatures. This is consistent with the work of McMahon, et.al. [4].

The physical property testing of all of the geotextiles immersed in water showed little change in strength and a general decrease in elongation relative to baseline data. This appears to be consistent with an increase in crystallinity within the amorphous region of the PET molecule.

Yet, when exposed to a highly alkaline environment, such as calcium hydroxide, at elevated temperatures the susceptibility of PET becomes much more striking.

In general, the staple fabric appeared to be more susceptible to degradation when exposed to various leachates and alkaline solutions. The continuous filament geotextiles did not lose strength after exposure to the landfill leachates, water or sodium hydroxide solution.

The continuous filament fabrics had an initial solution viscosity approximately 140 points higher than the staple fabric indicating a significantly higher molecular weight of the continuous filament geotextiles.

CONCLUSIONS

This study has demonstrated the relative durability of polyethylene terephthalate (PET) in various chemical environments.

Though too limited to demonstrate wide-spread chemical resistance, the data does clearly indicate that PET's resistance to leachates, chemical solutions, and water is related to the molecular weight of the polymer and that PET durability is an important concern in high pH environments.

REFERENCES

[1]   Risseeuw, P. and Schmidt, H. M, "Hydrolysis of HT polyester yarns in water at moderate temperatures", Proceeding of the 4th International Conference on Geotextiles, Geomembranes and Related Products, 1990, pp. 691-69.

[2]   Jailloux, J. M. and Verdu, J., "Kinetic models for the life prediction in PET hygrothermal aging: A critical survey", Proceedings of the 4th International Conference on Geotextiles, Geomembranes and Related Products, 1990, p. 727.

[3]   Cooke, T. F. and Rebenfeld, L., "Effect of Chemical Composition and Physical Structure of Geotextiles on Their Durability, Geotextiles and Geomembranes, Vol. 7 (1988), pp. 7-22.

[4]   McMahon, W., Birdsall, H. A., Johnson, G. R., and Camilli, C. T., "Degradation Studies of Polyethylene Terephthalate", Journal of Chemical Engineering Data, 1959, Vol. 4, No. 1, pp. 57-79.

ACKNOWLEDGEMENT

Portions of this paper are excerpted from a 1988 research report submitted to Hoechst Celanese Corporation by Texas Research Institute. Special thanks to Ms. Karen Verschoor and Mr. Rick Thomas, of Texas Research Institute, for the contributions as authors of the research report.

APPENDIX: QUALITATIVE CHARACTERIZATION of LEACHATES

CHEMICAL CONSTITUENTS OF LEACHATES ABOVE MINIMUM DETECTION LIMITS

	LEACHATE "L"	LEACHATE "N"	LEACHATE "O"
Aluminum	. . .	N/A	x
Antimony	. . .	N/A	x
Arsenic	x	N/A	x
Barium	x	N/A	x
Boron	x	N/A	. . .
Cadmium	x	N/A	x
Calcium	. . .	N/A	x
Chromium	x	N/A	x
Copper	x	N/A	x
Cyanide	. . .	N/A	x
Iron	. . .	N/A	x
Lead	x	N/A	x
Manganese	x	N/A	. . .
Mercury	x	N/A	. . .
Nickel	x	N/A	x
Potassium	. . .	N/A	x
Selenium	x	N/A	x
Silver	x	N/A	. . .
Sodium	. . .	N/A	x
Strontium	. . .	N/A	x
Vanadium	. . .	N/A	x
Zinc	x	N/A	x
Benzene	x	x	x
Chlorobenzene	x	x	x

Ethylbenzene	x	. . .	. . .
Toluene	x	x	. . .
Total Xylene	x	. . .	. . .
Methy Ethyl Ketone	. . .	. . .	x
Bis (2-Chloroethyl)		. . .	x
Ether	. . .	. . .	x
2-Nitrophenol	. . .	. . .	x
2,4-Dinitrophenol	. . .	. . .	x
M & P Cresols	. . .	. . .	x
4,6 Dinitro-O-Cresol	. . .	. . .	x
Acenaphthylene	. . .	. . .	x
Phenanthrene	. . .	. . .	x
Naphthalene	. . .	. . .	x
1,1-Dichloroethane	. . .	x	. . .
1,2-Dichloroethane	. . .	x	x
Methylene Chloride	. . .	x	x
Phenol	. . .	x	x
2,4-Dichlorophenol	. . .	x	. . .
Bis (2-ethylhexyl)			
phthalate	. . .	x	. . .
N-nitrosodiphenylamine	. . .	x	. . .
1,2,4-Trichlorobenzene	. . .	x	. . .

N/A: Not Available
. . .:   Not Present Above Minimum Detection Limits
x: Present Above Minimum Detection Limits

CHEMICAL CONSTITUENTS OF LEACHATES ABOVE MINIMUM DETECTION LIMITS

	LEACHATE "L"	LEACHATE "N"	LEACHATE "O"
Chloroform	. . .	x	. . .
1,2-Dichlorobenzene	. . .	x	. . .
Tetrachloroethylene	. . .	x	. . .
1,1,1-Trichloroethane	. . .	x	. . .
Trichloroethylene	. . .	x	. . .
pH (units)	6.9	N/A	N/A
Chloride (mg/L C_1)	10447	N/A	N/A
Sulfate (mg/L SO_4)	<25	N/A	N/A
Total Organic Carbon (mg/L C)	1,600	N/A	N/A
Total Dissolved Solids (mg/L)	17,900	N/A	N/A

N/A: Not Available
. . .:   Not Present Above Minimum Detection Limits
x: Present Above Minimum Detection Limits

Sam R. Allen and Karen L. Verschoor

MINIMIZATION OF THE EFFECT OF DIMENSIONAL
VARIABILITY ON GEOTEXTILE PROPERTY DATA FOR
INTERPRETATION OF CHEMICAL COMPATIBILITY TEST
RESULTS

REFERENCE: Allen, S.R., and Verschoor, K.L., "Minimization of the
Effect of Mass Dimensional Variability on Geotextile Property Data for
Interpretation of Chemical Compatibility Test Results", Symposium on
Geosynthetic Testing for Waste Containment Applications, ASTM STP
1081, Robert M. Koerner, editor, American Society for Testing and
Materials, Philadelphia, 1990.

ABSTRACT: Geotextiles have grown substantially, both in number
and in application, during the past ten years. The mechanical
properties of a specific region along a geotextile roll length are largely
dependent upon mass and dimensional characteristics corresponding to
that region. Geosynthetic manufacturers and testing laboratories have
adopted various sampling procedures and data manipulation schemes in
an effort to minimize the effect of product variability on mechanical
property test results. Alternative test methods have been employed that
demonstrate a high degree of repeatability. This paper reviews the
various approaches to the problem of dimensional variability and
discusses their rationale.

KEYWORDS: Geotextile, Normalization, Chemical Compatibility,
Variability.

Introduction and Background

Geotextiles are commonly used in hazardous and municipal waste
management facilities as critical components within liner and leachate
collection systems. The U.S. Environmental Protection Agency (EPA)
and state regulatory agencies are requiring the results of chemical
compatibility testing of geotextiles and other geosynthetic materials
used in a facility for permit applications. Results of chemical
compatibility testing are used to show the chemical resistance of
geotextiles with chemicals similar to ones that may be found in the
facility, and to examine durability as geotextiles perform their functions
of separation, reinforcement, filtration and drainage.

Mr. Allen is Project Coordinator at TRI Environmental, Inc., 9063 Bee Caves Road,
Austin, Texas 78733. Ms. Verschoor is the Program Manager of the Geosynthetics
Testing Program.

Although constructed of synthetic rather than natural materials, geotextiles are textiles in the traditional sense. Both demonstrate strength properties that are largely dependent on the individual component fibers as well as interactions between the fibers. Fiber characteristics are influenced by construction and process control parameters. Any change in these conditions can result in a considerable variation in a geotextile's weight, thickness and mechanical properties.

Several test methods currently used to characterize geotextile materials have been derived from similar test methods used in the textile industry. Many of the index properties measured to characterize geotextiles are used to characterize textile materials manufactured for the production of apparel. A geotextile's mechanical strength is directly related to its weight and thickness. This relationship must be understood and accounted for during analysis of a geotextile's strength properties.

Chemical compatibility studies involving index testing of geotextile material must take into account dimensional variability along the length and width of a geotextile roll since this testing depends on the ability to compare various mechanical properties before and after the material has been exposed to chemical media. The variability in chemical compatibility test results makes data interpretation difficult. An apparent change in mechnical strength could be caused by dimensional variations along a roll width usually consistent with a corresponding change in mass. Alternatively, the change may be due to actual degradation of the polymer. The first case may lead to false conclusions if, for example, a very thick section of geotextile were cut into samples and tested to generate baseline results, while samples at a later immersion interval were cut from a thin section.

Controlling the Effect of Dimensional Variability

Several procedures have been used to control dimensional variability for chemical compatibility studies involving index tests of geotextile mechanical properties. These include:

A.   Screening of test replicates for required weight and thickness;
B.   Sampling from a specific region across the width of a roll;
C.   Screening of test results to reject outliers;
D.   Tracking of test coupon dimensions for correlation to specific data and normalization of test results;

Method A minimizes the problem of dimensional variability by testing only pre-screened test specimens. Average roll values for weight and thickness are developed for a product roll dedicated to project testing by measuring and weighing a number of geotextile specimens (typically between 20 and 40) cut randomly from various regions of the material. Next, specimens for each index property to be tested at each exposure interval are cut from the roll. After cutting, each individual test specimen is weighed. If the weight falls outside one standard deviation from the established roll values, that specimen is rejected. Only test specimens falling within one standard deviation for both weight and thickness are accepted for project testing. Pre-screened specimens are not marked or identified, and weights are not recorded. The quantity to be cut for each index test is increased over the amount actually needed based on standard deviation values generated from average weight and thickness testing. A large number of extra coupons would need to be cut for geotextiles exhibiting a large degree of dimensional variability.

Method B recognizes that several geotextile manufacturing processes result in products in which weight and thickness vary uniformly across any given roll. In order to take advantage of this phenomenon, geotextile test specimens may be cut only from a selected region or width across the supply roll. Weight and thickness compliance with minimum average roll values is verified before testing begins. Again, specimens are not marked and corresponding dimensional data is not recorded. Choosing an appropriate region for cutting is critical to the success of this approach. Defining a region that is too wide may result in an undesirable degree of variability.

Method C involves screening of all test results before their inclusion into a data set. This is accomplished by testing a set number of replicates for each mechanical property analysis and rejecting a specified number of outliers so that the reported test value is generated from replicates exhibiting a lower variance. This method of reducing variability sometimes requires numerous test replicates and may result in data sets that continue to show excessive variability.

Methods A, B and C control dimensional variability based on the rationale that chemical compatibility testing is intended to determine whether a specific chemical media interacts with a geosynthetic product in any measurable way; since this issue is unrelated to product variability, the best approach is to eliminate this variability as much as possible, thereby simplifying data analysis.

Method D incorporates dimensional variability into the testing process by registering weights and thicknesses for each individual test specimen before subsequent exposure. This requires that a record be generated for all dimensional data collected for all test specimens to be tested. Each specimen is assigned a coded identification number and individual specimen weights and thicknesses are entered into a database which is kept for reference throughout the project. A numbered tag is attached to each specimen. Specimens are not screened; each weighed specimen is included in the test matrix regardless of weight or thickness. Prior to assigning specimens to exposure baths or intervals, all cut, weighed and tagged samples are "shuffled" physically so that weights and thicknesses throughout the range are evenly distributed. When test data are reported, original weights and thicknesses from the unexposed specimen database are included with the test results. Anomalous test values may be identified and considered in the context of any apparent trends in the data.

The tracking of individual coupon physical properties affords the testing laboratory the opportunity to incorporate dimensional characteristics into actual test results. The textile industry has used "normalization" of data to minimize the effect of dimensional variability in mechanical property test results. This may be accomplished by dividing the strength value generated from a test specimen by the corresponding mass per unit area or thickness for each coupon. A test coupon's weight and thickness may be used when both physical properties are related to the bulk strength. For example, a results expressed in force per unit area (e.g. lb/in^2) generated from a burst resistance test of a textile may be normalized by dividing the value by the mass per unit area (e.g. oz/yd^2) measurement for the corresponding test coupon, resulting in a "psi/osy", (pounds per square inch per ounce per square yard), an index of fabric strength used by the textile industry to reduce variability in test results caused by dimensional fluctuation along a roll width. In this same way, normalization may be employed in chemical compatibility studies involving a geotextile that exhibits a high degree of dimensional variability.

Determination of appropriate physical characteristics to use for data normalization is based on the specific test being analyzed. The result of a puncture test may be more related to a coupons thickness at the puncture location than the mass per unit area specific to the bulk sample. Mullen burst results may be related to both the thickness and the mass per unit area since the test region is enlarged and strength values are more dependent on bulk coupon properties. Before employing a normalizing formula, it must be established that a physical property is directly proportional to the specific strength property being measured. This may be accomplished by plotting the observed strength value for each unexposed test coupon vs. the corresponding physical property for that coupon. If a relationship is evident, normalization with that physical property measurement may be appropriate.

The success of data normalization may be assessed by calculating the relative coefficients of variation (v) resulting from raw and normalized data analysis.

The coefficient of variation is equal to the standard deviation divided by the mean for a particular population:

$$v = s/\bar{x}$$

where: v = coefficient of variation
          s = standard deviation
          $\bar{x}$ = mean

The coefficient of variation is a measure of data dispersion relative to the central tendency for a given data set. Properly normalized data sets should exhibit a reduced standard of deviation and a proportionately smaller coeffient of variation.

Table 1 gives puncture resistance values for a geotextile exposed to a chemical medium during a 120 day chemical compatibility study. Shown are averages for raw data generated at each thirty day time interval. In addition, averages calculated from normalized data sets are shown for comparison. Normalized puncture resistance results were generated by dividing observed force readings for each test coupon by the thickness measurement for that coupon. The coefficients of variance for each data set are given for each time interval before and after normalization of data. Finally, percent changes in observed puncture resistance are given for both raw and normalized data sets. Figure 1 graphically presents the effect of normalization of these data.

When using both the mass per unit area and thickness to normalize a mechanical property value, care must be taken to weight the two physical characteristics equally. This may be accomplished by using correction factors generated by defining the fractional relationship between each physical property and the average value of that property for an entire product roll. The resulting fractions may then be summed and divided by two to achieve an equal weight given to each. The resulting value may then be multiplied by the original strength property measurement to generate a "corrected" value. For example, a Mullen burst strength value in PSI for a specific test coupon might be normalized with the coupon's corresponding mass per unit area (oz/yd^2) and thickness (mils) by multiplying the original strength value by a correction factor generated as follows:

Correction Factor =

$$\frac{R(oz/yd^2)/C(oz/yd^2) + R(mils)/C(mils)}{2}$$

where: R represents product roll
            C represents individual test coupon

Table 2 gives the results of normalization of Mullen burst data collected during a geotextile chemical compatibility study. In this study, an observed Mullen burst strength value for a test coupon was found to be directly proportional to both the coupon's mass per unit area and thickness. For this reason, both physical properties were applied in the normalization process. The table includes coefficients of variance for each raw and normalized data set at each test period. Figure 2 shows test averages generated from these raw and normalized data sets. Normalized data points were calculated by multiplying each observed measurement by a correction factor generated as described above.

In both examples above, normalization serves to reduce noise in the data resulting from dimensional variability between the test coupons comprising each data set. The coefficients of variation for each data set were significantly reduced. While some variability still remains, the magnitude of observed changes in each mechanical property is more accurately assessed and trends may be more easily identified. The rationale for normalization is that an assessment of product variability can be obtained as a "by-product" of chemical compatibility testing, since weights and thicknesses are recorded and may be incorporated directly into the test result. In this way, the recorded dimensional data may prove useful in the evaluation of data, for example, to explain outliers or unexpected trends.

Problems with normalization occur when individual test coupons suffer physical property changes resulting from exposure to chemical media. Index property tests are typically performed using moist test coupons during chemical compatibility studies requiring that data sets be normalized with physical property values measured before exposure. If chemical exposure results in degradation of the polymer and subsequent loss of weight and thickness, normalization of test results with baseline physical property data becomes inappropriate. However, normalization may not be needed in this case since consistent loss in bulk physical properties usually result in related decrease of mechanical strength properties.

Control Groups

The effect of geotextile dimensional variability on mechanical test results may be checked by the use of a control group from which mechanical data is generated and compared statistically to data from other testing sets. Control sets typically consist of the specified geotextile exposed in deionized or tap water. Average results from mechanical tests on materials exposed in chemical media are compared against the corresponding average properties for the control specimens. Standard t-test distributions may be used to compare sample populations to the control group. The t-value may be computed for each mean value and related to a critical score corresponding to a specific confidence interval. For example, if the t-value exceeds 2.0 (a confidence interval equal to approximately 95 percent for a one-tail test, depending on the degrees of freedom), it may be concluded that the mean for the sample population differs statistically from the corresponding mean of the control group. Several statistical analyses may be applied with the addition of control group data to a geotextile chemical compatibility study, however a detailed discussion of these analyses is beyond the scope of this paper.

Error Bars

It is necessary to use error bars when presenting geotextile chemical compatibility results graphically. The inclusion of error bars provides perspective on observed changes in mechanical properties as a result of chemical exposure and enables a visual appreciation for product dimensional variability as it effects each data set during a chemical compatibility study. Presenting the data without error bars can give erroneous evidence of degradation.

The following graphs present puncture resistance data obtained during a chemical compatibility study. Figure 3 presents mean puncture resistance values for raw data given in Table 1 while Figure 4 shows the same data points with their corresponding error bars. The error bars shown represent ± two standard deviations from the mean, or a 95 percent confidence interval (assuming standard normal distribution).

Alternative Test Methods

EPA Method 9090 was developed in the early 1980's to give suggested procedures for testing the chemical compatibility of geomembranes. Geosynthetic design engineers and testing laboratories have adapted this test method to test geotextiles. Recently, geotextile chemical compatibility studies have included alternative mechanical tests that demonstrate a superior degree of repeatability when used to characterize geotextiles. For example, a strip tensile test is sometimes used to replace the traditional grab tensile test. The grab test utilizes a 4"x8" test coupon gripped by small 1 inch by 15 inch faces. It has been suggested that poor reproducicibly may be attributed to the large portion of specimen at the edges of the grips which is not contained during the test. A strip tensile test may be performed with a narrower sample coupon gripped along the full width of the specimen, thus reducing the possible influence of these effects.

Characterization of geotextile component fibers may also be employed to further avoid the compounding effects of end product variability. Characterization of fiber mechanical properties may be accomplished by performing tensile property testing in accordance with ASTM D2101 or ASTM D3822. Component yarns may be tested for tensile strength in accordance with ASTM D885. These tests often provide better precision for some geotextile products since the dimensional variability caused by the manufacturing process is effectively reduced. Currently, efforts are underway in the EPA and ASTM D35 Committee on Geotextiles, Geomembranes and Related Products to define and standardize protocols for chemical compatibility testing. Research is being performed that will result in a better understanding of the relative merits of geotextile index tests as they apply to chemical compatibility studies.

Summary

The procedures outlined above represent those most often used to control or minimize effects of geotextile dimensional variability on mechanical test results. The methods do not completely eliminate variability in test results and interpretation of chemical compatibility test data remains a complex task. However, because the results of chemical compatibility studies are reviewed by regulatory bodies to assess a material's capacity to perform its intended function in a specific application, a review of testing error introduced by dimensional variability is critical for proper interpretation of data. Using modified sampling procedures when preparing test coupons, recording physical properties for individual test samples, and data analysis using different statistical techniques such as normalization and control group comparison, have all been shown to reduce dispersion of test data caused by fluctuations in dimensional characteristics. Further experience with these methods and formulation of new techniques will serve to better define which approach should be used for specific projects.

TABLE 1: GEOTEXTILE PUNCTURE RESISTANCE

	0 DAY	30 DAY	60 DAY	90 DAY	120 DAY
AVG. PUNCT. RESISTANCE					
Raw (lbs):	253	168	161	234	178
Normalized (lbs/mil):	1.54	1.52	1.48	1.53	1.48
STANDARD DEVIATION					
Raw:	33	21	39	23	30
Normalized:	.15	.15	.27	.10	.15
% CHANGE IN PUNCTURE RESISTANCE:					
Raw:	- - -	-34	-36	-8	-30
Normalized:	- - -	-1	-4	-1	-4
COV					
Raw:	.13	.13	.24	.10	.17
Normalized:	.10	.10	.19	.06	.10
% CHANGE IN COV:	-23	-23	-21	-40	-41

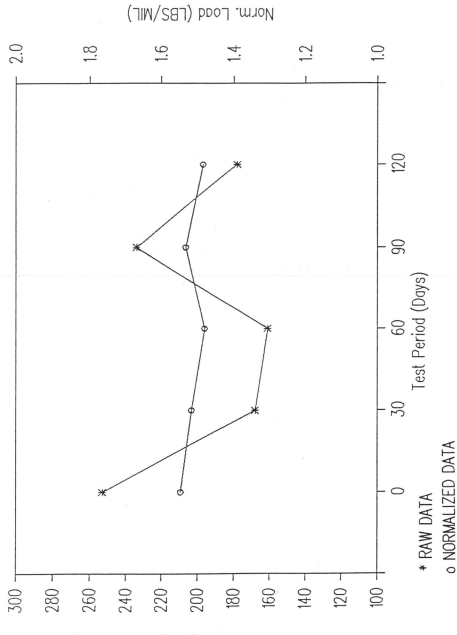

FIGURE 1: GEOTEXTILE PUNCTURE RESISTANCE

TABLE 2: GEOTEXTILE MULLEN BURST STRENGTH

	0 DAY	30 DAY	60 DAY	90 DAY	120 DAY
AVG. MULLEN BURST STRENGTH					
Raw (lbs):	155	162	146	168	132
Normalized (lbs/mil):	159	158	155	154	150
STANDARD DEVIATION					
Raw:	30	51	57	24	20
Normalized:	20	15	30	19	17
% CHANGE IN MULLEN BURST STRENGTH					
Raw:	---	-5	-6	-9	-15
Normalized:	---	-.50	-3	-3	-6
COV					
Raw:	.19	.31	.39	.14	.15
Normalized:	.10	.10	.19	.12	.11
% CHANGE IN COV:	-32	-68	-51	-14	-27

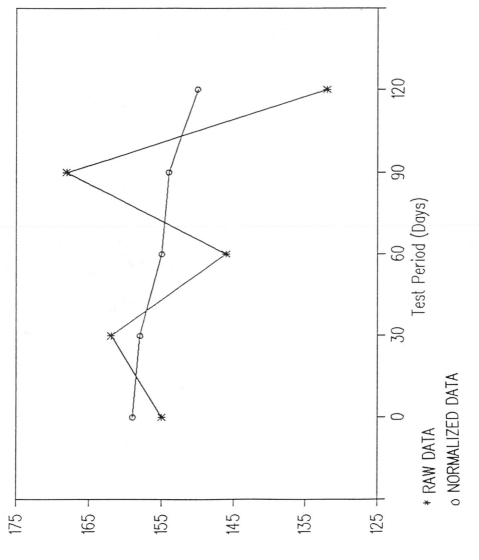

FIGURE 2:   GEOTEXTILE MULLEN BURST STRENGTH

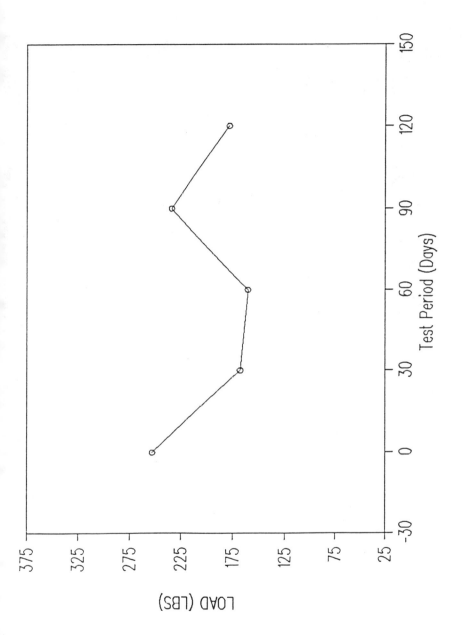

FIGURE 3:   PUNCTURE RESISTANCE (without error bars)

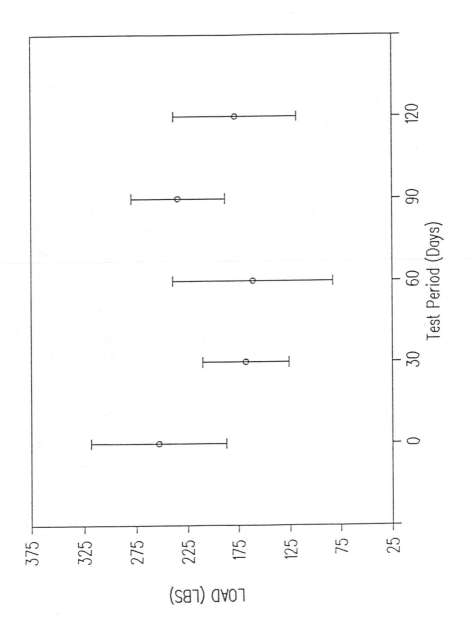

FIGURE 4: PUNCTURE RESISTANCE (including error bars)

Test Methods and Procedures to Evaluate Geotextiles and Other Geosynthetics

John A. Bove, P.E.

DIRECT SHEAR FRICTION TESTING FOR GEOSYNTHETICS IN WASTE CONTAINMENT

REFERENCE: Bove, J. A., "Direct Shear Friction Testing for
Geosynthetics in Waste Containment," Geosynthetic Testing
for Waste Containment Applications, ASTM STP 1081, Robert M.
Koerner, editor, American Society for Testing and Materials,
Philadelphia, 1990.

ABSTRACT: Sliding stability for landfills, heap leach pads
and caps constructed using geosynthetics is of critical
importance for both short and long term performance of these
facilities. For layered facilities, instability along
critical slip planes must be considered for construction and
operation of the facility.

The direct shear friction test, now being developed under
ASTM D35, is the test used to generate important performance
data for use in stability determinations.

This paper addresses the main elements of direct shear
friction testing and analysis including discussions on the
required test apparatus, specimen preparation and soil
placement. Differences in testing geosynthetics in contact
with other geosynthetics or soils are highlighted. Sources
of potential bias in test data are discussed along with
recommended solutions. Interpretation of direct shear
friction data for use in performance evaluation is included.

KEYWORDS: geosynthetics, geosynthetics testing, direct
shear testing, interface shear resistance, waste facility
design

Geosynthetics are an integral part of waste facility design,
construction, operations and closures. Their use has virtually
exploded due to The Resource Conservation and Recovery Act (RCRA) [1]
Subtitle C for hazardous waste facilities. Currently pending Subtitle
D regulations for municipal solid waste facilities also address the
application of geosynthetics. Clearly, the performance of

John A. Bove is the Geosynthetics Services Manager at Westinghouse
Environmental and Geotechnical Services, Inc., 11785 Highway Drive,
Suite 100, Cincinnati, Ohio 45241.

geosynthetics is of utmost importance in these facilities especially when the design life may approach one hundred years.

The primary elements of a waste facility where geosynthetics are featured are liner systems, leachate management systems (collection, removal or detection) and caps. The performance properties of geosynthetics within a landfill or surface impoundment may be classified as follows:

o Mechanical
o Hydraulic
o Chemical/Endurance

The specific properties and their relationship with geosynthetic performance are constantly being evaluated and are well documented [2-4]. This paper deals with the measurement of mechanical properties pertinent to the evaluation of sliding stability of elements within a waste facility. Specifically, the interface shear resistance of two dissimilar materials is addressed (which is really a combination of several mechanical properties that are conveniently treated as a single property).

SLIDING STABILITY

One significant advantage of the use of geosynthetics is the ability to construct liners, leachate management systems and caps on fairly steep slopes to maximize airspace. However, the installation of a geosynthetic on the slope of a landfill for instance, introduces a potential failure plane for sliding of overlying soils or waste. When not completely addressed during design, construction and operation, sliding failures can and do occur along these geosynthetic surfaces. Most failures can be prevented by proper testing, analysis and construction quality assurance.

Figure 1 is a sketch of a typical landfill cap cross section which shows a two foot thick cover soil overlying a geosynthetic drainage media (in this case a geotextile and a geonet) and the cap geomembrane. In order to evaluate the stability of the cap system, a block of cover soil is isolated using the free body diagram shown on Figure 2. The frictional resistance between the cover soil and the geotextile must be determined, but in order to completely analyze sliding stability, the following interfaces must also be considered:

o Geotextile to geonet
o Geonet to cap geomembrane
o Geomembrane to cap soil liner

The internal stability of the soil layers and the waste must also be verified. Similar analyses must be performed for landfill liner systems.

Generally, sliding failures on caps or landfill sideslopes are due to one or both of the following phenomena:

o Sliding along the weakest interface
o Tensile-type failure of the weakest geosynthetic component

In most applications, the required tensile resistance of a particular geosynthetic component is a direct function of the interface
friction between that component and the adjacent materials. The
greater the interface shear resistance of two surfaces, the greater
the capacity for transmitting shear to the underlying material. For
instance, the geotextile on the landfill sidewall shown on Figure 3
will have greater tensile force applied due to downdrag than will the
underlying smooth geomembrane. This is due to the shear resistance of
the geotextile/waste interface being greater than that of the
geotextile/geomembrane interface. If the geotextile/geomembrane
interface resistance is increased (perhaps by the use of a textured
geomembrane) the geomembrane will take on a proportionally greater
amount of tensile stress. For complex multiple-layer systems, it can
be shown analytically that the tensile stress applied to a particular
geosynthetic component is influenced more by the _difference_ between
the shear resistance of the upper and lower interfaces than the
magnitude of the normal stress itself. A more complete discussion of
this issue is presented in Richardson and Koerner [5], where the
relationship between system stability and interface shear resistance
is clearly seen.

EVALUATION OF INTERFACE FRICTION

Analysis for sliding stability of waste facilities is material-
specific and site-specific, and involves many factors that are outside
the scope of this paper. Measurement of interface resistance is a
critical component of stability analyses that can be performed in the
laboratory. A task group within ASTM D35 Committee on Geosynthetics
is in the process of establishing a standard test method for determining interface friction using the Direct Shear Method [6]. The main
elements of the test device, procedure and presentation of test data
are reviewed in this section.

TEST DEVICE

A schematic of a direct shear test device is shown on Figure 4.
The device consists of a pair of square or rectangular shear boxes
that each measure at least 300 mm (12 inches) wide and a minimum of 50
mm (2 inches) deep. One shear box is kept fixed while one is allowed
to travel horizontally under application of a shear force. A normal
compressive stress is placed on the specimen by means of a bellows,
weights or piston-applied force. The horizontal displacement of the
travelling shear box is measured, as well as vertical deformation, if
required.

Geosynthetic specimens are clamped to one or both of the shear
boxes in a way to constrain failure along a predetermined plane.
Typical clamping arrangements are shown in Figure 5.

The function of clamps in the direct shear test is to fix the
specimen(s) in order to allow failure at the predetermined interface.
Clamps must be designed and selected not to influence the interface
shear resistance of the specimens and to minimize (hopefully

eliminate) excessive tensile deformation outside of the specimen contact area.

The types of clamps used depend on the geosynthetic specimen and the specific direct shear equipment used. Clamps may be of the common wedge type fastened to one or both of the direct shear boxes. Pneumatic or hydraulic grips are effective for geonets and geogrids, especially where relatively high shear forces will be applied. An alternative to external clamps is to fasten the geosynthetic specimen to a rigid plate, as is discussed in more detail later in this paper.

A direct shear device for testing geosynthetics is generally quite large and made from steel and high performance bearing and loading systems. The equipment is not currently commercially available. The large device is intended to minimize end and scale effects that may be significant when testing some soil/geosynthetic or geosynthetic/geosynthetic systems in traditional soil direct shear devices (100 mm shear boxes). The 300 mm shear box allows for testing of larger gradation soils, such as gravel for leachate collection systems. It also enables the user to evaluate the mode of shear failure over a larger specimen area.

TEST PROCEDURE

The direct shear friction test method is described briefly in this section. Several steps in the test process are user specified and depend on the requirements of the particular project. The test is similar in concept to the traditional direct shear test for soils described in ASTM D3080 [7].

The interface shear resistance between a geosynthetic and soil or between a combination of geosynthetics is determined by placing the specimens in the test apparatus described above. A normal compressive stress is applied in a prescribed manner. The normal stress can be applied in a stepwise manner depending on the magnitude of the normal stress and the specimen conditions. A shear force is applied to the travelling shear box using a constant rate of displacement specified by the user. The applied shear force is recorded as a function of horizontal displacement of the travelling shear box. A typical plot is presented on Figure 6.

The user generally selects at least three different normal compressive stresses that best model anticipated field conditions. For each test run, the peak or residual shear value is selected and plotted against the normal stress applied for the test. A best fit straight line is used to connect the data points, as shown on Figure 7. The slope of this line, or failure envelope, is the coefficient of friction for the specimen interface where shearing occurred. This is typically converted to an angle, known as the angle of shearing resistance or the friction angle. The zero-normal stress intercept of the failure envelope is known as the adhesion. The friction angle and the adhesion are used to determine the interface shear resistance.

For waste containment applications, understanding the specific steps in the direct shear test is important. Since it is a performance test, test parameters are selected by the designer. Test data are quite sensitive to these parameters, especially for tests involving cohesive soils. Main elements of the direct shear test and test parameters are discussed in some detail below. The modes of specimen failure in the direct shear test are first presented.

Modes of Failure

The direct shear test measures shear resistance along a predeter-mined interface. This shear resistance may be a combination of many contributing factors, but the test method is not intended to distin-guish between individual factors. The components of a shear failure of the specimen can change with the applied normal stress, soil conditions and are sometimes a function of horizontal deflection of the travelling shear box.

Total shear resistance along an interface may be a result of one or a combination of the following factors:

o Sliding
o Adhesion
o Rolling of Soil Particles
o Interlocking of Soil Particles and Geosynthetic Surface
o Interlocking of Geosynthetic Surfaces
o Embedment of Soil Particles into Geosynthetic Surface
o Shear Strain of Geosynthetic

Identification of the failure mode and understanding the relationship between the mode of failure and the test parameters selected is important in understanding field performance of geosynthetics. The main elements of a direct shear test, are presented below along with discussions of the potential influence each element has on test results.

Specimen Preparation

Once the materials to be tested are selected, individual specimens are prepared. It is important to carefully select specimens that are as uniform as possible and are representative of the materials to be tested. This reduces the potential for biased test results that are caused by significant differences in the properties of the specimens. For instance, if a textured geomembrane is to be tested, each specimen should be cut in the same direction and contain the same degree of surface texture, which should also be representative of what will be used in the field. If one specimen contains less texture, the inter-face shear data for this test may not be comparable to data from other specimens. And more importantly, it may not be representative of actual field conditions. The general condition of the geosynthetic specimen, especially the shearing surface, should be carefully noted and recorded both before and after direct shear testing.

Soil may be placed in the lower shear box or above a geosynthetic specimen in the upper shear box. The moisture and density of the soil specimen are selected by the user and must be consistent for each test if accurate test data is to be recorded. Soil conditions, including the method of placement, have significant influence on the interface shear resistance between the soil and a geosynthetic specimen. Careful consideration of the soil conditions to be modeled in the laboratory and sound geotechnical practice in specimen preparation are necessary.

An important element in specimen preparation for direct shear testing is the selection of contact surfaces, referred to as the substratum and superstratum. Typical surfaces include rigid plates, a standard index surface such as sandpaper or soil. The substratum is used to support the geosynthetic specimen and to allow for uniform normal compressive stress distribution. The user must be aware of the potential effect of substratum or superstratum on test data. Consider the example of a geotextile tested against a soil. Placement of the geotextile specimen over a relatively smooth steel plate may yield an apparently lower interface shear resistance when tested with a single overlying layer of soil as opposed to being tested between two layers of the soil. In this instance, the use of the plate results in a smooth geotextile surface. Since the surface of the soil substratum is rarely this uniform after placement and compaction of the overlying soil layer, the substratum/geotextile interface may not be planar. This can result in increased interface shear resistance since failure is forced to be horizontal in the direct shear device. Although no soil surfaces constructed in the field are perfectly smooth, the effect of the laboratory specimens must be considered by the designer when evaluating direct shear test data.

Selection of Normal Compressive Stress

Normal compressive stress is selected to model anticipated field conditions as closely as possible. To do this, the designer must evaluate the anticipated normal field stress and "bracket" it by selecting at least one higher and one lower normal stress in addition to the anticipated value for direct shear testing. The performance of the test specimens over the entire range of normal stresses selected must be considered. Specimens may perform differently under shear failure at relatively low normal stresses, such as those representative in a landfill cap, than they would under higher normal stresses which would be present in a below waste liner system. As an example, consider a direct shear test of a needlepunched geotextile versus a compacted cohesive soil. At a low normal compressive stress of 6 kPa (140psf) the mode of failure at the soil/geotextile interface is predominantly sliding. Under a normal stress of 60kPa (1400 psf) the needlepunched geotextile, which has a relatively low tensile modulus, elongates significantly. The mode of failure is not only sliding, but tensile resistance of the geotextile must be overcome as well. The result is an apparently higher interface shear resistance at the higher normal stress. This brings to light two important points. First, the two sets of data described above should not be used to determine a single failure envelope for the specimen. This may result in an overestimation of interface shear resistance at low normal stresses, such as in cap applications. This is complicated further when one considers that the failure envelope may not be a true straight line at very low confining pressures. For these reasons, extrapolation of the failure envelope beyond the test data points is not recommended for use in design of waste facilities.

It is recommended that the designer specify additional direct shear tests if the above conditions are encountered. Additional tests can be clustered around the extreme sets of direct shear data in order to identify the approximate normal stress where interface behavior apparently changes. For a specific application, the designer may option to reduce the range of normal compressive stresses selected to more closely model anticipated field conditions.

A second point concerns the use of published angles of interface friction for a variety of specimen combinations. The designer should be cautious in selecting such performance parameters unless he confirms that the test specimens and test conditions are applicable for his specific use. For design of waste facilities, however, this practice is not recommended.

Deformation Rate

The proposed direct shear test method recommends the following deformation rates for tests where none is specified by the designer:

Geosynthetic vs. Geosynthetic: 5 mm/min (0.2 in/min)
Geosynthetic vs. Soil: 1 mm/min (0.04 in/min)

These guidelines are intended to allow interlaboratory reproduction of direct shear test data. The user should select the deformation rate to best match the field application. For geosynthetic vs. geosynthetic tests tested in the dry, the interface shear resistance is generally not sensitive to deformation rate if the rate is less than approximately 12 mm/min (0.5 in/min).

For tests involving soils, the type of soil and soil conditions (density, moisture content, etc.) strongly influences test results. The large size of the test device may preclude running a truly drained test for cohesive soils, even if the thickness of the soil specimen is reduced. For these tests, a deformation rate much less than 1 mm/min may be required if a drained test is desired. In many tests for waste containment, geomembranes can prevent soil drainage. For the reasons listed above, direct shear tests involving cohesive soil and geosynthetics are generally performed under undrained conditions.

Inspection of Specimen

After completion of each direct shear test, it is important to carefully remove and inspect the test specimen, including that portion within the clamps. The condition of the specimen can provide the designer insight into the mode of failure. This is important. If the failure mode is identified, the user can gain a greater understanding of the test results and nonrepresentative data can be discarded. Such information allows for adjustment in the test procedure if required. The following is a partial list of items that should be checked and recorded for geosynthetic specimens after each test and during the test if possible:

o Inspect geomembrane surfaces for abrasion and any pattern of abrasion, etc.
o Signs of elongation (uniform or localized?) or other damage
o Development of wrinkles
o Disbonding of geocomposites
o Embedment of soil particles in geosynthetic surface
o Differential movement between specimen and contact surface

o Excessive deformation at clamps
o Tilting ("racking") of platens or other horizontal surfaces.

For some testing programs, it is desirable to return geosynthetic specimens along with test data for evaluation by the designer.

Interpretation of Results

For each test, a plot of shear stress versus horizontal deflection similar to the one shown on Figure 6 is generated. This plot provides a good indication of how the specimen behaved during shearing. Usually the designer looks for a distinct peak and residual behavior from the curve.

The peak stress represents the stress required to overcome the combination of friction, adhesion and passive resistance to initiate sliding along the interface. The residual stress is the stress required to maintain sliding and is less than or equal to the peak stress. The failure envelope (friction angle and adhesion) may be determined from either the peak or residual points on the shear stress versus deformation plots. The residual friction angle may be equal to the peak stress, but is generally lower.

The shape of the shear stress versus deflection plot may be as significant as the magnitude of peak or residual shear stress. When plotted during the test, this curve gives a quick indication of the specimen failure mode.

The plot presented on Figure 6 represents a classical sliding-type mode of interface shear failure. This is what would typically be expected when testing a HDPE geomembrane against compacted clay or against a geotextile at relatively low normal stresses. For shear failures that occur predominantly in the sliding mode, the distinct peak normally occurs within the first 10 mm (0.5 inch) of horizontal displacement of the shear box. If there is no distinct peak in the shear stress versus horizontal displacement plot, or if the peak occurs at a greater displacement, other models of shear failure may be occurring along with sliding. Several examples are presented on Figure 8 (a) - (d) and are described below.

Figure 8(a) is a shear stress versus horizontal displacement plot that exhibits no distinct peak, but does exhibit residual behavior. This shape of curve is indicative of shear strain within the geosynthetic specimen, typically low tensile modulus materials such as a needlepunched geotextile or a PVC geomembrane. In this case, a portion of the applied stress is used to elongate the geosynthetic specimen. As the specimen elongates, its tensile modulus increases, gradually allowing shear transfer to the interface. For this example, peak and residual angles of friction are essentially equal. The designer must determine if this type of behavior (that is, significant elongation of the geosynthetic) is anticipated in the actual waste facility, and whether this elongation is tolerable.

The type of response shown on Figure 8(a) may occur for either soil/geosynthetic or geosynthetic/geosynthetic tests. One additional example where this may occur is a lightweight needlepunched geotextile against a textured HDPE geomembrane.

One method to reduce elongation of relatively low modulus specimens is to attach the specimen, using glue or other methods, to a rigid substrate. This can allow a sliding-type failure to occur and the resulting direct shear data may be compared to the unconstrained case.

For some cohesionless soils, such as sands or gravels, "rolling" of individual particles may occur during direct shear testing. When tested against a surface that provides a good interface bond, such as a nonwoven geotextile or textured geomembrane, pure sliding will not occur at the soil/geosynthetic interface. Shear stress applied to the specimen may be resisted by the interlocking of such particles during "rolling" of loose sand or gravel. In this instance, a shear stress versus horizontal deflection plot similar to that shown on Figure 8(b) may result. The shear stress never peaks since the soil particles have not been "locked" together to the degree necessary to transfer shear to the geosynthetic interface. It is difficult to determine the angle of friction for a series of tests that exhibit this behavior. It should be noted that this is more a soil shear failure than a true soil/geosynthetic interface failure. When this is the case (assuming this type of failure is not acceptable to the designer) additional soil compaction may be needed, or a soil with different gradation or particle shape may be substituted.

A shear stress versus horizontal displacement curve like the one shown on Figure 8(c) exhibits no true residual shear behavior. This type of response can occur when a portion of a geosynthetic specimen becomes imbedded in a soil substratum. It can also occur when testing geonet/geocomposites against soil at fairly high normal compressive stresses. The multiple peaks can coincide with the passing of the geonet nodes over a localized high area of the soil specimen.

Finally Figure 8(d) presents an example of one of the many combinations of failure modes that can occur during a single direct shear test. The initial (lower) peak can represent sliding of the soil/geosynthetic interface. At a horizontal deformation of approximately 8 mm for this example, the applied shear stress causes the geosynthetic specimen to elongate. At a deformation of 30 mm the shear is again transferred back to the specimen interface. The user may choose to ignore the highest peak value and select the more conservative initial peak for design.

The above discussion does not address all potential responses. It does reinforce the need to present the designer not only with values of peak or residual angles of friction and adhesion, but with complete test data as well.

CONCLUSIONS

Sliding stability of soil masses is a critical step in the design of waste containment facilities. Stability concerns become more acute when soils are combined with geosynthetics. The direct shear friction test, currently being evaluated by ASTM Committee D35 on Geosynthetics, can aid the designer in determining the interface shear resistance of soils and geosynthetics or a combination of geosynthetics. Test parameters such as specimen placement, normal compressive stress, and shearing rates must be   selected by the designer to best   model field conditions.

Laboratory testing must be conducted under conditions that model anticipated field conditions as closely as possible. The designer must first attempt to identify field conditions for each facility and each design element that may significantly impact geosynthetics and soils performance. Laboratory testing and material selection can then be completed using anticipated conditions to satisfy design requirements.

The modes of failure of the specimen must be understood by the designer. The failure response of a particular specimen may be understood by evaluating the shear stress versus horizontal deflection plot for the test.   This can provide the designer with additional insight to predict field behavior as well as allow him to identify inappropriate test data.

REFERENCES

[1] USEPA, "Hazardous Waste Management System; Permitting Requirements for Land Disposal Facilities," Part II, Federal Register Washington, D.C., pp. 32274-32388 (July 26, 1982).

[2] Koerner, R.M., Designing With Geosynthetics, Prentice-Hall, New Jersey (1986).

[3] Geotextile Testing and the Design Engineer, ASTM STP 952, Fluet, J.E. Jr., ed., American Society for Testing and Materials, Philadelphia, 1987.

[4] Proceedings: Geosynthetics Conference '89 Industrial Fabrics Association International, San Diego, CA, February, 1989.

[5] USEPA "Geosynthetic Design Guidance for Hazardous Waste Landfill Cells and Surface Impoundments," EPA/600/S2-87/097, Richardson, G.N., and Koerner, R.M., Hazardous Waste Engineering Research Laboratory, Cincinnati, OH, February, 1988.

[6] ASTM Committee D35 on Geosynthetics Subcommittee D35.01 on Mechanical Properties.

[7] "Standard Test Method for Direct Shear Test of Soils Under Consolidated Drained Conditions" ASTM D3080, Annual Book of ASTM Standards, Vol. 04.08, American Society for Testing and Materials, Philadelphia, PA.

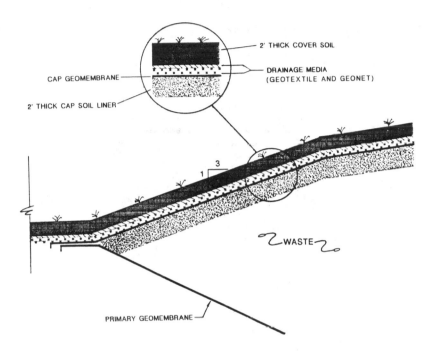

Figure 1 - Sketch of Landfill Cap Cross Section for Sliding Stability
        Analyses

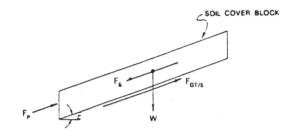

F_P = Resisting Force of Passive Block
F_S = Driving Component of Soil Block Mass
W = Normal Component of Soil Block Mass
$F_{GT/S}$ = Frictional Force Between Soil Block and Geotextile

where β = Slope Angle
δ = Angle of Friction Between Soil and Geotextile
       Determined from Direct Shear Test

Figure 2 - Free Body Diagram of Soil Block for Sliding Stability Analyses

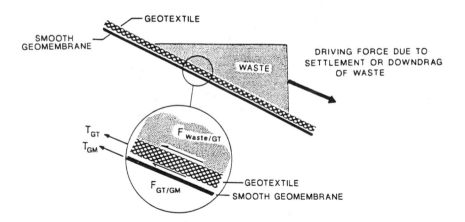

Figure 3 - Schematic of Resisting Forces to Sliding of a Waste Mass

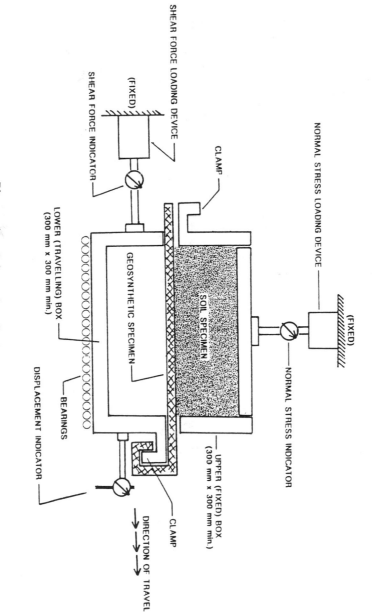

Figure 4 - Schematic of Direct Shear Device [6]

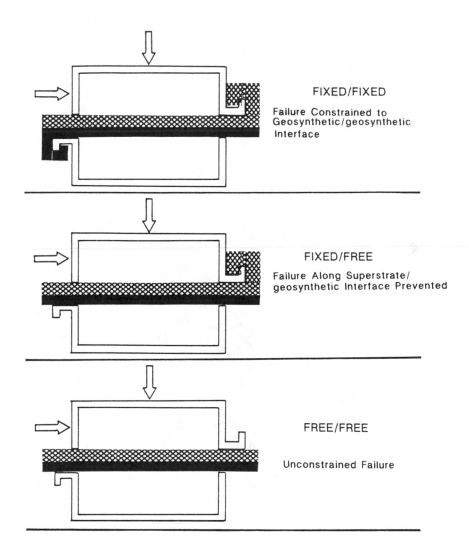

Figure 5 - Typical Clamping Arrangements for Geosynthetic/Geosynthetic
Direct Shear Testing [6]

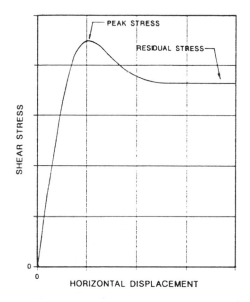

Figure 6 - Typical Plot of Shear Stress versus Horizontal Displacement
of Direct Shear Test

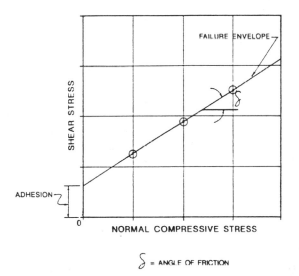

Figure 7 - Typical Plot of Shear Stress versus Applied Normal
Compressive Stress for Direct Shear Test

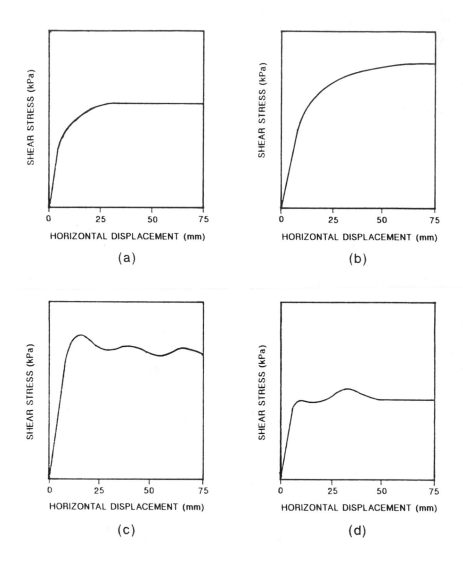

Figure 8 - Typical Shear Stress versus Horizontal Displacement for
          Several Potential Direct Shear Interface Failure Modes

          (a)   Plot with no Distinct Peak and a Flat Residual
                Behavior
          (b)   Plot with no Peak nor Residual Behavior
          (c)   Plot with Distinct Peak but no Uniform Residual
                Behavior
          (d)   Plot with Two Peaks Representing Bi-Modal Failure

Hoe I. Ling[1], Fumio Tatsuoka[2], and Jonathan T. H. Wu[3]

MEASURING IN-PLANE HYDRAULIC CONDUCTIVITY OF GEOTEXTILES

REFERENCE: Ling, H. I., Tatsuoka, F., and Wu, J. T. H., "Measuring In-Plane Hydraulic Conductivity of Geotextiles," Geosynthetic Testing for Waste Containment Applications, ASTM STP 1081, Robert M. Koerner, editor, American Society for Testing and Materials, Philadelphia, 1990.

ABSTRACT: An apparatus was developed to measure the in-plane hydraulic conductivity of geotextiles. The apparatus is capable of measuring geotextile transmissivities under specified constant hydraulic heads, subject to simulated stress conditions, and in the confinement of soil. Using the new apparatus, a nonwoven geotextile and a woven-nonwoven composite geotextile were tested to measure their in-plane hydraulic conductivities under various normal stresses. Excellent repeatability of the test results were obtained. Three means of confinement for the geotextile (namely, using rigid blocks, flexible membranes, and soil) were employed for simulation of the in-soil condition. In addition, geotextile specimens retrieved from a field embankment were tested to examine the changes in its transmissivities from those of fresh specimens.

KEYWORDS: test apparatus, nonwoven geotextile, composite geotextile, in-plane hydraulic conductivity, transmissivity, soil confinement, field extracted specimen

Geotextiles serves two important hydraulic functions related to cross-plane flow (filtration) and in-plane flow (in-plane drainage) when embedded in soil. There are various experimental approaches to studying these properties and the literature has been summarized in reference [1]. Some thick geotextiles which have sufficient interconnecting pore structures to allow fluid flow within its plane

1) Graduate Student of Civil Engineering, Univ. of Tokyo, Japan.
2) Associate Professor, Institute of Industrial Science, Univ. of Tokyo, Japan.
3) Associate Professor, University of Colorado at Denver, U. S. A. (currently Visiting Professor, Univ. of Tokyo, Japan)

257

have found increasing popularity in in-plane drainage applications. However, relatively less study has been done on the in-plane hydraulic conductivity of geotextiles when compared to that of the cross-plane. Koerner, et al. [2] identified numerous common applications in which the in-plane flow capacity of a geotextile is of an important design consideration. These applications include chimney drains in dams, drains behind retaining walls, and pore water dissipators in cohesive earth fills.

In the Tokyo area, five test embankments of volcanic ash clay, reinforced either with a non-woven or a woven-nonwoven composite geotextile, have been constructed. The performance of these test embankments has been described in detail in references [3, 4, 5, 6]. The height of the embankments ranged from 4.0 m to 5.5 m with slopes of 1.0 : 0.05, 0.2, 0.3 (vertical : horizontal). It was found that other than reinforcing the embankment wall, the geotextile sheets also improved compaction by draining pore water from the interior of the embankments.

The test methods commonly used for measuring the in-plane hydraulic conductivity of geotextiles are (1) the parallel flow test, suggested by the ASTM Committee on Geotextiles, Geomembranes, and Related Products (D 4716-87)[7], and (2) the radial flow test [2]. In the parallel flow test, fluid flow occurs along a longitudinal flow path in which the geotextile specimen is placed so that stream lines of flow through the geotextile are generally in a parallel trajectory. Using a circular disk geotextile specimen, the radial flow test allows flow to enter the geotextile specimen at the inner circumference. The stream lines of the flow therefore radiate from the center of the circular disk outward in all directions. These tests are designed to determine geotextile transmissivity under specified constant hydraulic head conditions and under varying normal stresses. In both tests, the normal stresses on the geotextile specimen are applied through a rigid plate by an external loading mechanism such as static weights.

In this study, a new test apparatus for measuring in-plane hydraulic conductivity of geotextiles was developed. Some features of the test apparatus are listed below:
(1) It is a constant head parallel flow permeameter.
(2) The operational stress condition in field installation can be simulated.
(3) Different means of confinement, including using soil as the confinement, can be employed.
(4) Using soil confinement, time dependency (clogging/blocking) of the hydraulic conductivity in the geotextile can be investigated.
(5) The apparatus can also be adapted to measure cross-plane hydraulic conductivity of geotextiles.

Using the developed test apparatus, two geotextiles, a nonwoven geotextile and a woven-nonwoven composite geotextile, which were used in the above-mentioned test embankments were tested to measure their in-plane hydraulic conductivities under various normal stresses. Three different means of confinement (using rigid blocks, flexible membranes, and soil) were used to examine the effect of using other confinements for simulation of the in-soil condition. The in-soil tests were repeated to examine the reproducibility of the test procedure. In addition, geotextile specimens retrieved from a clay embankment two and half years after their installation were tested to examine the changes in transmissivities from those of the fresh specimens.

TEST APPARATUS

Figure 1 depicts the configuration of the parallel-flow type constant head test apparatus used in this study. The apparatus is composed of three major components:

Water supply tank: The total heads at the water supply and water receiving tanks, TH_1 and TH_2, respectively, were maintained constant. This was achieved by applying a constant air pressure p_1 through the tip of pipe ② inside the water supply tank ①. Meanwhile, the temperature of the water was measured with a thermometer ③ inside this supplying tank.

Water receiving tank: Water flowed through the geotextile specimen before emerging from another tube ⑤ erected in the water receiving tank ④. A different constant air pressure p_2, in this case the atmospheric air pressure, was applied to the receiving tank. The total discharge was collected in this tank and measured with a low-capacity differential pressure transducer (LC-DPT) ⑥.

Permeameter: The geotextile specimen ⑨ was housed in a triaxial cell ⑦ modified for the purpose of this study. Two water dispensing meshes ⑧ were placed at the two ends of the specimen to achieve a uniform parallel flow condition. The head loss between the two ends of the specimen was measured with another LC-DPT ⑩. The actual head loss in the specimen was obtained by subtracting the apparatus frictional and shape head losses from the measured head loss. The details on the calibration of apparatus head loss were described in References [1, 8].

The effective stress in the specimen was measured with a high capacity differential pressure transducer (HC-DPT) ⑪. An axial load was applied through the loading piston ⑬, and the thickness of the specimen was measured using a displacement transducer ⑫ in cross-plane flow tests. The tubing ⑭ joining the top cap was coiled to follow the movement of the loading cap. Cell pressure was applied by air ⑮, and if required, a deviator load could be applied on the top plate ⑯.

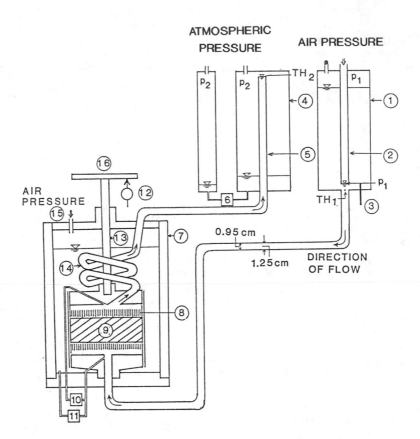

FIG 1 -- The constant head permeameter.

It has to be noted that the test apparatus has also been used to measure the cross-plane hydraulic conductivity of geotextiles, the results of which are reported in references [1, 8].

TEST MATERIALS

Geotextiles

Two types of fresh geotextiles were used in this study. One was a spun-bonded nonwoven geotextile made from polypropylene fibers. The other was a composite geotextile with a layer of woven geotextile, 0.5-mm thick, interbedded between two thin layers of nonwoven geotextile. Index properties of the geotextiles provided by the manufacturer are shown in Table 1.

TABLE 1 -- Index properties of the geotextiles

Properties	Nonwoven geotextile	Composite geotextile
unit weight (gf/m²)	460	310
thickness (mm)	4	3
tensile strength (kgf/5cm)	110 (70)	90 (15)
elongation at break (%)	100 (120)	n.a.
puncture strength (kgf)	30 (35)	10 (18)

the number in the ( ) indicates the value for cross machine direction.

Soil

For tests performed using soil as confinement, Kanto loam was used. This clay is a volcanic ash clay with clay, silt, and sand contents of 29 percent, 63 percent, and 8 percent, respectively. The particle size distribution curve of Kanto loam is shown in Figure 2. In the tests, the soil was prepared at a water content of about 100 percent, which is similar to the value in the test embankments described earlier.

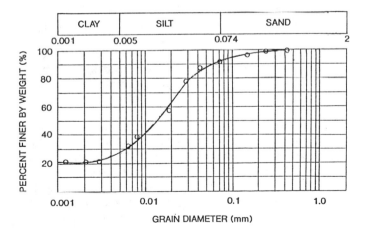

FIG 2 -- The particle size distribution curve of Kanto loam clay.

TESTING CONDITION

Three different methods of confinement were selected for simulation of the confinement of geotextiles in their operational condition. These confinement methods are as follows:

Method 1: Geotextiles confined with rigid acrylic blocks;
Method 2: Geotextiles confined with flexible membranes; and
Method 3: Geotextiles confined with soil, in this case Kanto loam.

For Method 1, the geotextile specimen was simply sandwiched between two rigid acrylic blocks over the entire planar area of the specimen and subjected to different confining pressures.

For Method 2, the geotextile specimen was confined by two flexible membranes, one on each side, and confining pressures were applied through two pieces of flexible material (e.g. clay) onto the specimen. This method was used in order to obtain better contact between the confining material and the geotextile surface.

For Method 3, i.e., the in-soil tests, two thin layers of compacted Kantoloam, approximately 1-cm thick, were placed on each side of the geotextile specimen. The soil-geotextile composite was then confined by two pieces of rigid acrylic blocks as shown in Figure 3. The soil-geotextile composite along with the acrylic blocks was in turn confined inside a rubber membrane sealed to the cap and pedestal of the modified triaxial cell. Great precautions were taken to ensure that no air was trapped in the sample and no prestress was inadvertently applied to it. One piece of the block was used to support the pedestals and cap and was restrainted from any movement whereas the other was made 3 mm shorter to allow for free horizontal movement upon the application of confining pressures. As the confining pressure was applied, the shorter piece of the acrylic blocks moved inward and transferred the pressure to the geotextile specimen. It is to be noted that the soil merely served to confine the geotextile and the small thickness of clay was used in order to minimize consolidation of the soil.

The hydraulic conductivities of Kanto loam and the geotextile were approximately in a ratio of 1 to 10,000. Consequently, for the thickness of the soil and geotextile used in this study, the discharge contributed by the clay was only 10^{-6} times the total discharge. This implies that the soil can be considered impermeable in the soil-geotextile composite. For the in-soil tests, pieces of geotextile were attached to the ends of the soil blocks to prevent the soil from being eroded.

For Method 1, the transmissivities in both the machine and cross machine directions were measured for both geotextiles. For Methods 2 and 3, all the transmissivities were measured in the machine direction.

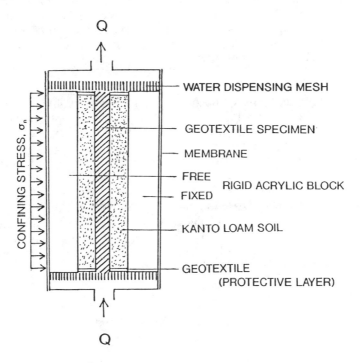

FIG 3 -- Confinement condition of the in-soil tests.
(Not to scale)

Effective confining pressures in the range of 0.1 to 1.5 kgf/cm^2 were selected for this study. The range of hydraulic gradient used was in the range of 0.025 to 3.

Approximately ten sets of successive readings of physical quantities, viz. temperature of the water, compression of the specimen, effective stress and head loss in the specimen, and the discharge were recorded periodically at one-minute intervals at low stress levels and larger intervals (more than five minutes) at high stress levels through an analog-to-digital converter using a microcomputer . An averaged value was used for each individual quantity.

FIELD EXTRACTED GEOTEXTILE SAMPLES

Among the five large-scale clay test embankments reinforced with geotextile sheets, three of them were at the Experimental Station of the Institute of Industrial Science, University of Tokyo. The second test embankment, subsequently referred to as Embankment II, was constructed of Kanto loam in the early 1983 with a nonwoven geotextile of similar properties to the nonwoven geotextile described earlier but with a nominal thickness of 3 mm.

Long-term monitoring had shown that Embankment II performed satisfactorily, in terms of both the mechanical and hydraulic behavior in the reinforced zone [3, 4, 5]. For example, the embankment was able to maintain a relatively high suction during heavy rainfalls in the course of the monitoring program, which implied that the water which infiltrated might have been drained by the upper layers of the geotextile. Embankment II was demolished after two and half years monitoring using artificial rainfall. The exposed cross-section after the rainfall test is shown in Figure 4. Numerous geotextile specimens were retrieved from the embankment upon its demolition and stored under air-dry condition. Three of the specimens were selected for testing in their machine direction using the in-soil confinement with Kanto loam after soaking them in a rubber membrane.

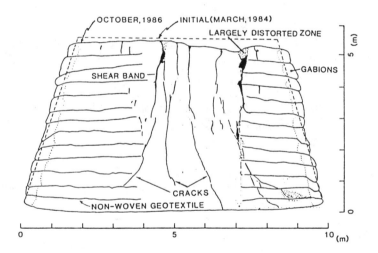

FIG 4 -- Cross-section of Embankment II during dismolition.

DATA INTERPRETATION

Hydraulic conductivity in the plane of nonwoven geotextile is characterized by the coefficient of in-plane permeability, k_h or transmissivity, θ. Following Darcy's law,

$$k_h = \frac{q}{A}\frac{L}{h} \qquad (L/T) \qquad (1)$$

and

$$\theta = \frac{q}{W}\frac{L}{h} \qquad (L^2/T) \qquad (2)$$

where q is the rate of discharge, $A(=Wxt_g)$ is the cross sectional area of flow, t_g is the geotextile thickness, h is the total head loss in the geotextile specimen, W and L are, respectively, the width and length of the specimen in the flow direction. In this study, W and L were about 11 cm and 8 cm, respectively. A constant temperature of 20 °C was maintained during the tests.

For computing the in-plane permeability, k_h, the change in thickness of the geotextile with normal stress was determined by a hyperbolic strain-stress relationship [1, 8], shown in Figure 5,

$$\varepsilon_n = \frac{\sigma_n'}{a + b\,\sigma_n'}, \qquad (\%) \qquad (3)$$

where a and b are the geotextile material constants which can be determined by a compression test. The thickness t_g of the geotextile at any effective confining stress is determined by $t_g = (1-0.01\,\varepsilon_n)t_{g0}$, where t_{g0} is the initial thickness of the specimen determined prior to each test.

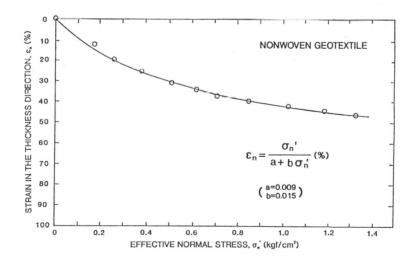

FIG 5 -- Compressibility of the nonwoven geotextile.

RESULTS AND DISCUSSIONS

Figure 6 shows the relationships of total head and flow velocity of the nonwoven geotextile tested under different confining pressures. Linear constitutive relationships between flow velocity and total head loss were obtained for all the confining pressures, as indicated by Darcy's Law. Linear constitutive relationships were also observed in the tests with geotextiles confined by rigid blocks and flexible membranes. It is of interest to note that a nonlinear constitutive relation with a treshold hydraulic gradient was observed for the same geotextile when flow occurred in the cross-plane direction [1, 8].

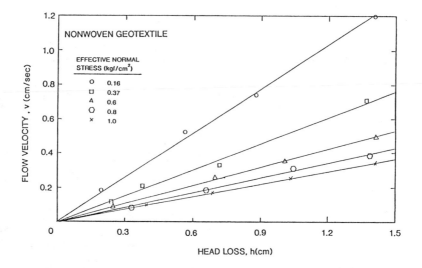

FIG 6 -- Relationship between flow velocity and total head loss of the nonwoven geotextile.

Figure 7 shows the relationship between transmissivity and effective normal stress for the nonwoven geotextile in both the machine and cross-machine directions. As may be expected, the hydraulic conductivity reduces with increasing effective normal stress. It is seen that the geotextile exhibits a slight degree of anisotropy and that the in-plane hydraulic conductivity is strongly dependent upon the effective normal stress, especially at low stress levels (up to approximately 0.6 kgf/cm^2). Similar behavior was observed for the composite geotextile.

Figure 8 shows the results under the three methods of confinement (using rigid blocks, flexible membranes, and soil). It is seen that the

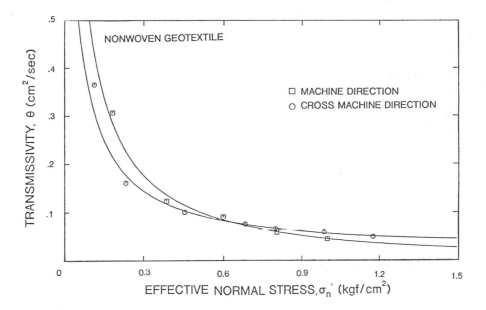

FIG 7 -- Relationship between transmissivity and effective normal stress
of the nonwoven geotextile.

in-plane hydraulic conductivity for specimens confined with rigid
acrylic blocks was higher than that confined with flexible membranes.
The in-soil permeability was the lowest among the three confinement
conditions. The differences are due to the different interface
conditions between the geotextile specimen and the confining materials.
The flow rate was larger for a larger interface gap formed between the
geotextile surface and a stiffer contacting material. Besides, the soil
particles were seen to have penetrated into the geotextile matrix and
resulted in about a 10 percent increase in the unit weight after drying.
The same results were obtained for the composite geotextiles, as shown
in Figure 9.

Repeatability of the in-soil test was also investigated. Three
in-soil tests were performed with the same confining soil blocks.
Consequently, in the second and third tests the soil blocks had been
prestressed and might have been of somewhat different densities from
one another. From Figure 10, it is obvious that the in-soil test results
are highly reproducible. The density of the soil appeared to have
little effect on the measured hydraulic conductivity if the soil used is
of a small thickness. However, the effect of using different soil types
warrants further research.

It is seen from Figures 8 and 9 that the nonwoven geotextile has a
higher hydraulic conductivity than the composite geotextile under the

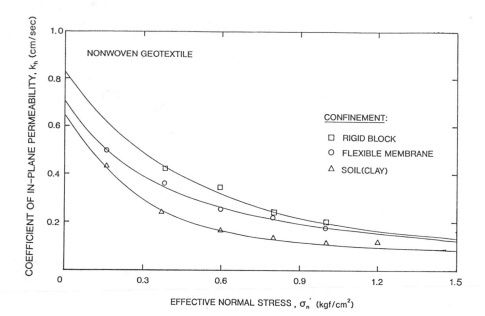

FIG 8 -- Coefficient of in-plane permeability of the nonwoven geotextile under different confinement conditions.

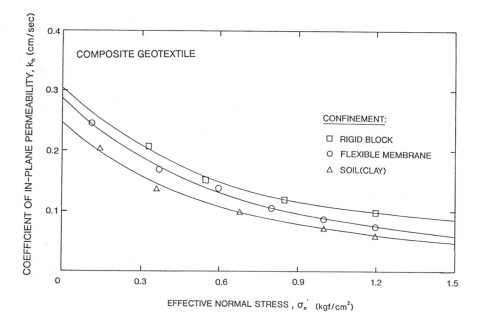

FIG 9 -- Coefficient of in-plane permeability of the composite geotextile under different confinement conditions.

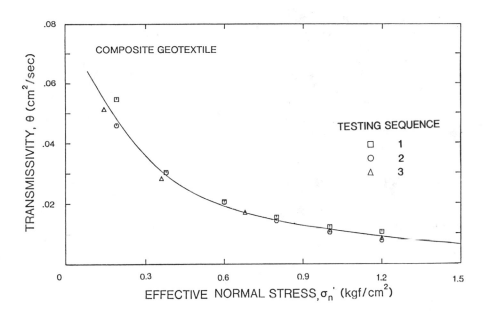

FIG 10 -- Repeatability of the in-soil tests.

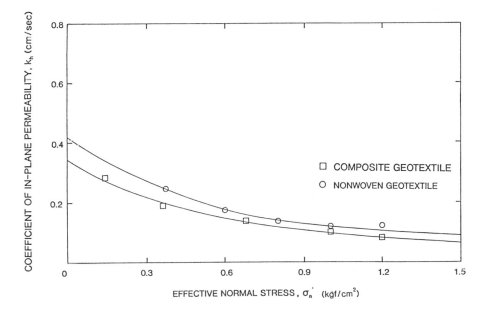

FIG 11 -- Comparison of the coefficient of permeability of the nonwoven
and composite geotextiles.

same confinement condition. However, this difference is due primarily to the thickness of the nonwoven geotextile. By comparing their in-plane permeabilities without considering the woven geotextile inclusion in the geotextile composite, the two geotextiles have very similar flow characteristics as shown in Figure 11. This implies that the woven inclusion in the composite geotextile conducts little flow.

In Figure 12, the transmissivities of the fresh and the extracted nonwoven geotextiles are shown. A reduction of the transmissivity, up to about 75 percent, was obtained for the geotextile situated close to the base of the embankment. The primary reason for the reduction of the transmissivity is believed to be due to clogging of the geotextile by fine soil particles. However, the reduction did not seem to have significantly affected the performance of the embankment as its hydraulic conductivity was still much higher than that of the embankment fill.

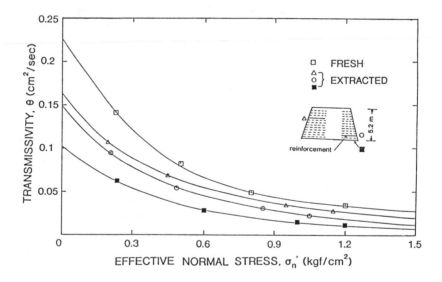

FIG 12 -- Relationship between transmissivity and effective normal stress of the extracted nonwoven geotextile.

CONCLUSIONS

From this study, it is concluded that

i) The apparatus developed in this study can be used to measure accurately in-plane hydraulic conductivity of geotextiles in the confinement of soil. From the test results, it is shown that the in-soil tests were very repeatable.

ii) The in-soil test simulates the operational condition of geotextile in the most realistic manner while confinement with rigid blocks or flexible membranes tend to overestimate the in-situ hydraulic conductivity. For the two geotextiles used in this study, rigid block confinement gave as much as two times higher conductivities than using soil confinement. Their ratios were greater at higher stress levels.

iii) Tests performed on geotextile specimens retrieved from a field embankment indicated significant reduction in the transmissivity two and half years after their installation. However, the reduction appeared not to have affected the hydraulic performance of the earth structure.

iv) Woven inclusion in the composite geotextile does not appear to play a role in its in-plane drainage function.

ACKNOWLEDGMENTS

The first author would like to express his appreciation and gratitude to the Japanese government for providing the financial assistance for his Master of Engineering study at the University of Tokyo. Moreover, the authors would like to acknowledge Mr. J. Nishimura of the Mitsui Petrochemical Industries, Ltd., Mr. T. Sato, Mr. T. Ishii, and all the members of the Geotechnical Laboratory for their kind assistance in this study.

REFERENCES

[1] H. I. Ling, "Measuring hydraulic and mechanical properties of geotextiles under soil-confinement conditions," M. Eng. Thesis, University of Tokyo, Japan, Feb., 1990.

[2] R. M. Koerner, J. A. Bove and J. P. Martin, "Water and air transmissivity of geotextiles," Geotextiles & Geomembranes, Vol. 1, No.1, 1984, pp. 57-73.

[3] F. Tatsuoka and H. Yamauchi, "A reinforcing method for steep clay slopes with a nonwoven geotextile, "Geotextile & Geomembrane, Vol. 4, 1986, pp. 241-268.

[4] H. Yamauchi, F. Tatsuoka, K. Nakamura, Y. Tamura, and K. Iwasaki, "Stability of steep clay embankments reinforced with a non-woven geotextile," Proc. of Post Vienna on Geotextiles, Singapore, pp.370-386, 1987.

[5] F. Tatsuoka, K. Nakamura, K. Iwasaki, Y. Tamura, and H. Yamauchi, "Behavior of steep clay embankments reinforced with a non-woven geotextile having various face structures, "Proc. of Post Vienna on Geotextiles, Singapore, pp.387-403, 1987.

[6] F. Tatsuoka, O. Murata, and M. Tateyama, Discussion, Int. Conf. on Soil Mech. and Found. Engrg, Rio de Janeiro, Vol. 4, pp.1311-1314, 1989.

[7] ASTM Standard D 4716-87, "Standard test methods for Constant head hydraulic transmissivity (in-plane flow) of geotextiles and geotextile related products," Vol. 04.08, pp.874-877, 1989.

[8] H. I .Ling, J. Nishimura and F. Tatsuoka, "Flow normal to the plane of a nonwoven geotextile and formulation of flow equation," Proc. of 44th Annual Convention of Japanese Society of Civil Engineers, Nagoya, pp. 1034-1037, Oct., 1989. (in Japanese)

Claudia M. Montero and Leo K. Overmann

GEOTEXTILE FILTRATION PERFORMANCE TEST

REFERENCE: Montero, C. M., Overmann, L. K., "Geotextile Filtration Performance Test" Geosynthetic Testing for Waste Containment Applications, ASTM STP 1081, Robert M. Koerner, editor, American Society for Testing and Materials, Philadelphia, 1990.

ABSTRACT: A rapid test has been developed to observe the actual filtration performance of nonwoven geotextiles with site-specific soils. The test is a variation of the "Slurry Test" developed by Dr. James L. Sherard, and is performed by replacing the granular medium in Sherards' test with a geotextile. Several tests were performed with fine-grained soils and nonwoven geotextiles of different mass per unit area, thickness and apparent opening size (AOS). It was observed that the phenomena that occurred in the geotextile filtration test are similar to described by Sherard during his research. In successful tests, the initial outflow is turbid but clears within a short period of time and a "filter cake" forms at the geotextile-slurry interface. In unsuccessful tests, a large amount of soil passes through the geotextile and/or no "filter cake" forms.

KEYWORDS: geotextile, nonwoven, needlepunched, heatbonded, continuous filament, apparent opening size (AOS), filtration, fine-grained soil

This paper describes the background, procedures, results, analyses and conclusions associated with a testing program developed to observe and evaluate the filtration performance of several nonwoven, continuous filament geotextiles with

Ms. Montero is a Senior Engineer and Mr. Overmann an Associate with Golder Associates Inc., 3730 Chamblee Tucker Road, Atlanta, Georgia 30341.

fine-grained soils.

A series of twelve fine-grained soil filtration tests were performed with nonwoven geotextiles of different manufacturing processes, mass per unit area, thickness and apparent opening size (AOS). The AOS values of the tested materials ranged from 0.600 mm (0.0236 in.) for a 0.23 mm (0.009 in.) thick, 65 g/m^2 (1.9 oz/yd^2) heatbonded geotextile to 0.090 mm (0.0035 in.) for a 5.33 mm (0.210 in.) thick, 550 g/m^2 (16.2 oz/yd^2) needlepunched geotextile. The testing procedure developed is performed rapidly and easily, and simulates conservative filtration conditions.

The physical phenomena of filtration that occurred between the fine-grained soils and the geotextiles in this testing program is comparable to the results of similar testing previously performed with conventional granular filters. In successful tests, the soil is retained as a "filter cake" created at the contact surface of the filter medium. In unsuccessful trials, the soil passes directly through the filter medium.

SIGNIFICANCE OF TEST

Due to the increasing cost of waste disposal volume, engineers involved with the waste industry frequently elect to utilize synthetic materials as impermeable barrier, drainage and filtration layers in place of their more conventional, thicker, natural component counterparts. As a result, the application and use of the various geosynthetic materials is increasing rapidly. The specific application of these materials varies with a number of factors, including: facility location, size, and depth; availability of construction materials; number, type and relative location of the various components, and; overall complexity of the design.

In a modern waste containment unit, compacted, fine-grained, relatively impermeable soil barriers are occasionally placed in direct contact with geotextile components. An example of this is a dual liner and leachate collection/removal system in which a geonet and geotextile drainage layer is directly overlain by a compacted fine-grained soil layer.

In the role of designer, the authors have become aware of the intricacies involved in specifying a geotextile filter which will perform in direct contact with fine-grained soils. There are a number of existing design criteria offering guidance to the specification of geotextiles for the filtration of soils [1,2]. However, the majority of these criteria apply to coarse-grained soils. The criteria that do apply to fine-grained soils are generally conservative [2]. The conservative nature of the existing design criteria has lead to the development of this performance-based approach.

BACKGROUND

The geotextile filtration performance test employed in the testing program described herein is a variation of a procedure developed by Dr. James L. Sherard in 1981-82 [3]. Dr. Sherard investigated critical downstream granular filters for fine-grained soils in central core, earth dams. Sherard demonstrated that such a situation could be modeled in the laboratory by applying a large hydraulic gradient to a slurry made of fine-grained soil, placed over a granular filter.

Sherard observed that in successful filtration tests a soil "filter cake" develops at the contact between the granular filter and the slurry. Typically, a small amount of turbid water passes through the sample during the first few minutes of the test. Subsequently, the outflow is clear. In unsuccessful tests, all of the slurry is forced through the filter in a short period of time, the upper surface of the filter is left clean, and no soil "filter cake" is formed.

It is unlikely that such high hydraulic gradients would develop within the confines of a waste containment unit. Nevertheless, the performance of a geotextile filter should be similar to that of a granular filter. Therefore, by using a geotextile filter in place of a granular one, Sherards' slurry test method can be used to observe the performance with fine-grained soils.

DESCRIPTION OF TEST SETUP

Three fine-grained soils were selected for testing and individually mixed with water to form a slurry. The slurries were placed on a sample of candidate geotextile secured at the bottom of a soil permeameter mold as shown in Fig. 1. A hydrostatic pressure was then applied to the top of the slurry and the outflow observed. The results of the test were subsequently evaluated based on the soil content of the initial outflow and on the ability of the geotextile to promote the creation of a soil "filter cake" at the geotextile-slurry contact surface. The details of the test apparatus, materials, and procedures are described as follows.

Description of Slurry Soils

Samples of three fine-grained, low permeability soil barrier layer materials from waste disposal facilities with landfill units under construction were obtained for use in this testing program. The particle size distributions and index properties of these soils are shown on Fig. 2.

Soil Types A and B were obtained from the same disposal facility site and illustrate the variability of properties

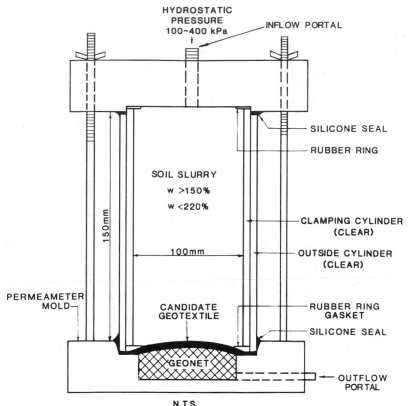

FIG. 1 -- Geotextile filtration test setup.

that might be expected in a borrow source.   Soil Type C was obtained from a different disposal facility.

The particle size distribution and Atterberg limits of each soil were determined before the slurries were prepared. Each of the soils is classified as a CL material in accordance with the Unified Soil Classification System (USCS).   In all cases, at least 65 percent of the soil particles are finer than 0.075 mm (0.0030 in.).   These soils, particularly soil Type B, are considered to be some of the finest that would typically be encountered in impermeable soil barrier layer applications.   The liquid limit (LL) and plasticity index (PI) of the three soils range from 38 to 48 and from 19 to 32, respectively.

Each of the soils was mixed with tap water, allowed to soak for approximately 24 hours and mixed, by hand, to the approximate consistency of automobile motor oil. A dispersing agent was used to facilitate the mixing process of two of the twelve slurries utilized.   The slurries were determined to have moisture contents ranging from approximately 150 to 220 percent.   This moisture content range was selected based upon

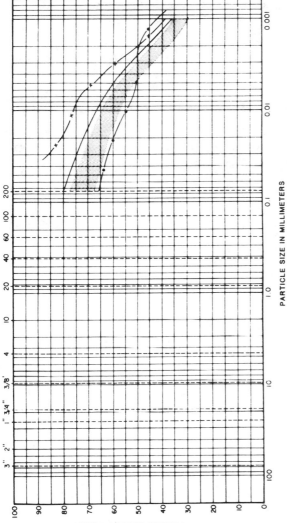

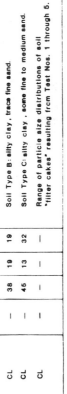

COBBLES	GRAVEL		SAND			FINES	
	COARSE	FINE	COARSE	MEDIUM	FINE	SILT OR CLAY	

CURVE	U.S.C.S CLASSIFICATION	Wₙ	LL	PL	PI	DESCRIPTION OR CLASSIFICATION
——	CL	—	48	21	27	Soil Type A: silty clay , some fine sand.
— x —	CL	—	38	19	19	Soil Type B: silty clay , trace fine sand.
— • —	CL	—	45	13	32	Soil Type C: silty clay , some fine to medium sand.
▨	CL	—	—	—	—	Range of particle size distributions of soil "filter cakes" resulting from Test Nos. 1 through 5.

FIG. 2 -- Particle size distribution curves.

the previously described work performed with granular filters and fine-grained soils.

Description of Geotextile Materials

A total of twelve laboratory tests were performed on six different geotextile materials. Nine tests were performed on nonwoven, needlepunched, continuous filament, polyester geotextiles of four different mass per unit area, produced by a single manufacturer. Three additional tests were performed on two different mass per unit area, nonwoven, heatbonded (on both sides), continuous filament, polypropylene geotextiles produced by a single, but different manufacturer.

The mass per unit area of the nonwoven, needlepunched geotextiles tested were distributed as follows: three 550 g/m^2 (16.2 oz/yd^2), three 360 g/m^2 (10.5 oz/yd^2), one 200 g/m^2 (6.0 oz/yd^2) and two 120 g/m^2 (3.4 oz/yd^2). According to the manufacturers' literature, the AOS of these materials ranges from 0.090 to 0.212 mm (0.0035 to 0.0083 in.). These values encompass the typical range of nonwoven, needlepunched geotextiles currently produced in the United States.

Two of the three tests conducted on nonwoven, heatbonded geotextiles involved 65 g/m^2 (1.9 oz/yd^2) materials with a reported AOS of between 0.425 and 0.600 mm (0.0167 and 0.0236 in.). The third test was conducted on a 115 g/m^2 (3.4 oz/yd^2) geotextile with a reported AOS of between 0.212 and 0.250 mm (0.0083 and 0.0098 in.).

Testing Apparatus

The geotextile filtration apparatus employed in this testing program consisted of a 100 mm (4 in.) diameter standard soil permeameter, slightly modified to accommodate concentric, clear, plastic cylinders, approximately 150 mm (6 in.) in height. A cross-section of the apparatus is shown in Fig. 1.

The inner cylinder, resting on flat rubber ring gaskets at both ends, is utilized to clamp the candidate geotextile in place at the base of the apparatus. The outer, or sealing cylinder is caulked around the circumference of both ends with silicone to form a water/airtight seal. The inset at the base of the apparatus is filled with two layers of geonet material to accommodate the potential flow volume through the overlying candidate geotextile to the outflow portal.

At the time of this writing an apparatus was being fabricated which will eliminate the need for the silicone sealant and therefore facilitate assembly.

GEOTEXTILE FILTRATION TESTING

Procedure

Each of the geotextile filtration performance tests was set up as described above and as shown in Fig. 1. The soil slurry was carefully poured directly onto the surface of the exposed candidate geotextile at the base of the inner cylinder, to a level of approximately 125 mm (5 in.). The remainder of the cylinder height was then filled with tap water in a manner to exclude air from the apparatus.

The inflow valve was connected to a water or air column and a pressure ranging from 100 to 400 kPa (15 to 60 lbf/in.^{2}) was applied. The outflow was measured for 60 seconds and at periodic time intervals thereafter.

The volume and moisture content of the initial discharge was determined. The tests were continued until the outflow became clear to the unaided eye. If a leak developed in the silicone seal the test was immediately stopped.

At the completion of a test, the apparatus was disassembled and the slurry remaining in the cylinder was carefully poured out. The thickness of the soil "filter cake", if one had formed, was measured manually. In addition, the soil "filter cake" was carefully removed from the surface of some geotextiles and subjected to a particle size distribution analysis.

Although the intent of the testing program was to observe only the initial filtration performance of various geotextiles, results for longer test durations were also obtained. Five of the tests were executed for periods of up to 90 hours. It was necessary to periodically supplement the slurry with additional water throughout the course of these longer duration tests. The longer duration tests were generally terminated when leaks formed in the silicone seals.

Results

The interpretation of the test results is based on the success of the candidate geotextile in retaining the majority of the soil in the slurry. As a general rule, the presence of a soil "filter cake" at the contact surface between the geotextile and the slurry indicated successful filtration performance.

A comparison of the moisture content of the slurry to that of the outflow obtained during the first 60 seconds of testing provides a measure of the quantity of soil solids per unit volume of water filtered and passed through the candidate geotextile. These quantities are an indication of the effectiveness of the geotextile to perform as a filter for the fine-grained soil. Typical test results obtained during this

TABLE 1 - Summary of geotextile filtration test results.

Test No.	Geotextile Properties			Soil Type[c]	Water or Air Pressure (kPa)	Slurry Moisture Content (%)	60 Sec. Outflow Moisture Content (%)	60 Sec. Outflow Volume (ml)	Percent of Solids Retained On and In Geotextile After 60 Seconds	Percent of Solids Passing Through Geotextile After 60 Seconds	Soil "Filter Cake" Formed (Y/N)	Results and Remarks
	AOS[a] (mm)	Type[b]	Mass Per Unit Area (g/m²)									
1	0.090-0.125	NP	550	A	100	180	1958	8	90.8	9.2	Yes	Test passed.
2	0.090-0.125	NP	550	A	200	201	16848	51	98.9	1.2	Yes	Test passed.
3	0.090-0.125	NP	550	A	200	221	5868	10	96.2	3.8	Yes	Test passed.
4	0.106-0.150	NP	360	A	200	177	1520	11	88.4	11.6	Yes	Test passed.
5	0.106-0.150	NP	360	A	200	177	4489	56	96.1	3.9	Yes	Test passed
6	0.106-0.150	NP	360	C	400	151	16260	50	99.1	0.9	Yes	Test passed.
7	0.150-0.212	NP	200	C	100	178	17610	70	99.0	1.0	Yes	Test passed.
8	0.150-0.212	NP	120	C	100	151	15207	67	99.0	1.0	Yes	Test passed.
9	0.150-0.212	NP	120	B	100	178	608	105	70.7	29.3	Yes	Inconclusive results.
10	0.212-0.250	HB	115	B	200	151	213	60	29.1	70.9	Yes	Probable Failure. Clogged after 1 min.
11	0.425-0.600	HB	65	B	200	178	178	All	0.0	100.0	No	Test failed.
12	0.425-0.600	HB	65	C	200	151	178	110	15.2	84.8	No	Test failed. No outflow after 1 min. Leak in silicone seal developed.

[a] Per manufacturers' literature.

[b] NP = Nonwoven, continuous filament, needlepunched geotextile.
HB = Nonwoven, continuous filament, heatbonded geotextile.

[c] Particle size distribution curves and index properties of soil Types A, B, and C are shown in Fig. 2.

testing program are summarized in Table 1.

The most common feature of all successful tests performed was that the initial outflow, regardless of its volume or moisture content, was always very turbid for a short period of time and subsequently became clear. The duration of turbid outflow ranged from 10 to 40 seconds.

No definite pattern in outflow volume or moisture content was observed. Additionally, the results appear to be independent of the applied water/air pressure with the exception of minor variations in the duration of turbid outflow.

Needlepunched Geotextiles

All tests performed using needlepunched geotextiles retained the majority of the soil contained in the slurry and eventually allowed clear water to flow through. The initial outflow from each of the tests was turbid and contained a variable amount of soil solids. The solids content of the outflow ranged from approximately 1 to 30 percent of the initial weight of soil per unit volume of water in the slurry. The outflow became clear as a soil "filter cake" formed above the geotextile.

Particle size distribution analyses were run on several of the soil "filter cakes" recovered from these tests and were invariably found to be coarser than the original slurry soil. The particle size distribution curves of the tested soil "filter cakes" are shown on Fig. 2.

Each of the three needlepunched geotextiles used in the first seven tests of the program formed soil "filter cakes" and successfully filtered the two soils tested. The AOS of these geotextiles ranged between 0.150 and 0.212 mm (0.0059 and 0.0083 in.) for the material used in Test No. 7 to between 0.090 and 0.125 mm (0.0035 and 0.0049 in.) for the material used in Test Nos. 1 through 3.

The needlepunched geotextile used in Test No. 8 caused a "filter cake" to form when performed with Soil Type C. Although a "filter cake" was also formed when the same geotextile was tested with the finer soil in Test No. 9, the relatively large volume of soil solids in the outflow prove this material inconclusive to function as an adequate filter.

Heatbonded Geotextiles

The testing program originally conceived was intended to evaluate the filtration performance of only nonwoven, needlepunched geotextiles with fine-grained soils. However, geotextiles of this type are generally only available in the United States with an AOS of 0.212 mm (0.0083 in.) or less. Because the results of the tests performed with the needlepunched geotextiles near the upper limit of AOS values available were somewhat inconclusive, the program was continued with another geotextile type. This additional testing required the use of nonwoven, heatbonded materials. Their use also provides an opportunity for the comparison of the filtration abilities of different types of nonwoven geotextiles.

The results of these additional tests are reported as Test Nos. 10 through 12 in Table 1. These results indicate that the heatbonded geotextiles with an AOS of 0.212 mm (0.0083 in.) or more allowed greater than 70 percent of the soil solids in the slurry to pass through during the first minute of testing, regardless of the soil type.

Although a soil "filter cake" formed at the geotextile-slurry interface in Test No. 10, the heatbonded material used failed to retain enough soil solids to be considered successful. The geotextiles used in Test Nos. 11 and 12, with an AOS of between 0.425 and 0.600 mm (0.0167 and 0.0236 in.), failed to cause a soil "filter cake" to form and/or allowed relatively large volumes of slurry to pass through. In Test No. 11, which involved the use of soil Type B, the entire contents of the cylinder passed through the geotextile in approximately 20 seconds.

INTERPRETATION OF RESULTS

Although the soil with the finest particle size distribution was not used in testing either of the two needlepunched geotextiles with an AOS of 0.150 mm (0.0059 in.) or less, it is anticipated that these materials will likely function as a filter for most, if not all, fine-grained soils commonly found in nature. On the other hand, needlepunched geotextiles with an AOS between 0.150 and 0.212 mm (0.0059 and 0.0083 in.) will adequately function as a filter for only certain fine-grained soils. In this program, these geotextiles filtered soils with less than 65 percent passing the 0.075 mm (0.0029 in.), as represented by Soil Type A. However, additional testing is required to determine the performance of these materials with finer soils.

The single heatbonded geotextile with an AOS of between 0.212 and 0.250 mm (0.0083 and 0.0098 in.) appears inadequate to filter soils with 100 percent finer than 0.075 mm (0.0029 in.). Additional testing is required to determine more definitively the limitations of this geotextile material.

The results of the testing performed on the two heatbonded geotextiles with an AOS of 0.425 to 0.600 mm (0.0167 to 0.0236 in.) are conclusive as to the inadequacy of this material to function as a filtration layer for the soils tested in this program.

SUMMARY AND CONCLUSIONS

This test method is rapidly and easily performed and, through direct observation, is useful in the qualitative determination of the filtration performance of specific soils and geotextile products. However, the method remains somewhat inconclusive to the formulation of broad generalizations regarding the development of empirical design criteria for the filtration performance of nonwoven geotextiles and fine-grained soils. The formulation of such design criteria would require additional testing with a wide range of soils, including silts and highly plastic clays, and geotextiles, both needlepunched and heatbonded.

As a performance-based test, this method provides no definitive means of assessing the safety factor associated with selection of a geotextile for a specific application. However, the method does provide a relative means of comparing "acceptable" geotextiles to one another and to "unacceptable" ones.

The use of slurried soils in this test method provides an inherent safety factor relative to the state of the soil in the condition of actual use. The slurried condition of the soil provides the soil solids more mobility than would otherwise be realized. This safety factor may be quantified if additional testing were performed to compare the results of a testing program such as this to another performed with the same soils in other conditions (ex: loose, compacted, etc.).

In conclusion, the slurry test method, as described in this program, is a viable laboratory method of evaluating the filtration performance of specific nonwoven geotextile and fine-grained soil systems. However, the testing program that has been performed to date is insufficient to conclusively determine trends to formulate general design criteria. Additionally, with minor modifications to the apparatus and/or procedures, this test could be used to evaluate additional design parameters.

RECOMMENDATIONS FOR ADDITIONAL TESTING

The described testing program was developed to investigate only the filtration performance of geotextiles and fine-grained soils. However, with minor modifications of the testing procedure, additional geotextile design criteria parameters may be investigated.

The clogging potential and permeability of a specific geotextile-soil system could be determined with a procedure similar to that described in this program. Through time, the soil used in the slurry of this test will tend to settle from suspension. As it does, the outflow from the system should approach a rate which is equivalent to that produced through the mass of consolidated soil solids. If the geotextile is clogged with soil to the extent that it is ineffective or if the geotextile is not sufficiently permeable, it is expected that the outflow would decrease to a level of less than the steady-state rate through the soil mass.

This geotextile filtration performance test method could easily be modified to analyze dynamic and/or reverse flow conditions. These conditions could be modeled by cyclic application of hydrostatic pressure and reversal of the inlet and outlet portal, respectively or in combination, as may be necessary by the desired conditions. With minor modifications of the apparatus, it may also be possible to model the filtration characteristics of geotextiles under uniformly loaded conditions.

REFERENCES

[1]   Giroud, J. P., "Designing with Geotextiles," in Geotextiles and Geomembranes: Definitions, Properties and Design, Third Edition, Industrial Fabrics Association International, St. Paul, Minnesota, 1985, pp. 263-270.
[2]   Geotextile Engineering Manual Course Text, prepared for the Federal Highway Administration, National Highway Institute, Washington D.C., 1984.
[3]   Sherard, J. L., "Filters for Silts and Clays," Journal of Geotechnical Engineering, ASCE Volume No.110, No.6, June 1984, pp. 701-710.

Shobha K. Bhatia, Saad Qureshi and Robert M. Kogler

LONG-TERM CLOGGING BEHAVIOR OF NON-WOVEN GEOTEXTILES WITH SILTY AND GAP-GRADED SANDS

REFERENCE: Bhatia, S.K., Qureshi, S., and Kogler, R.M., "Long-Term Clogging Behavior of Non-Woven Geotextiles with Silty and Gap-Graded Sands," Geosynthetic Testing for Waste Containment Applications, ASTM STP 1081, Robert M. Koerner, editor, American Society of Testing Materials, Philadelphia, 1990.

ABSTRACT: Non-woven geotextiles are generally used for the protection of synthetic liners in sanitary landfills. In this application the geotextiles perform as a filter and drain for the leachate collection system. Since geotextiles play a major role in landfill design, they should be properly selected to obviate any clogging problems. Filtration criterion for the non-woven geotextiles are still evolving. The existing filter criteria are not applicable for silty and gap-graded soils. In this paper, the results are presented for six different non-woven geotextiles which are tested with eighteen different gap-graded soils to evaluate long-term filtration behaviors. Microstructure analysis results are provided for a better understanding of the clogging mechanism of non-woven geotextiles.

KEY WORDS: Non-woven geotextiles, long-term filtration, clogging, microstructure, silty and gap-graded soils.

INTRODUCTION

The purpose of a leachate collection system (LCS) in a waste landfill is to efficiently collect the generated liquid, drain it to a down gradient sump area and rapidly remove it for proper treatment and disposal. Furthermore, such collection systems must be kept free flowing for the entire service life and post closure period of the facility. If the leachate collection system does clog, the accumulated liquid will either find a hole in the liner system and be forced through it by an ever increasing hydraulic head, or eventually diffuse through the liner. In either case, the negative implications toward subsurface contamination are obvious [1].

To ensure effective operation of the leachate collection system, the designer must ensure that the system does not become clogged due to bacterial clogging or accumulation of soil particles. Specific attention should be given to the horizontal boundaries such as between the LCS and adjacent soil or waste deposits, and around the collector pipe network within the LCS.

Dr. Bhatia is an Associate Professor at the Department of Civil and Environmental Engineering, Syracuse University, Syracuse, NY 13244-1190; Saad Qureshi is working with Woodward Clyde Consultants, Gaithersburg, MD 20878; and Mark Kogler is a graduate assistant at the Department of Civil and Environmental Engineering, Syracuse University, Syracuse, NY 13244-1190.

In these boundaries often non-woven geotextiles are used as a filter system. Thus, it is imperative to insure that the selected geotextiles will perform as an effective filter.

In order to use the geotextiles as a filter material, they must satisfy three criterion: Soil retention, permeability and clogging. These filtration criterion for non-woven geotextiles are still evolving. The soil retention criterion for non-woven geotextiles are generally based on an apparent opening size (AOS) for the geotextiles. The AOS of the geotextile is usually evaluated in the laboratory using a test procedure developed by the U.S. Army Corps of Engineers [2]. This test measures the percent of uniform glass beads retained on the geotextile for a range of bead sizes. The bead size having only 95% retained is defined as the O_{95} or AOS of the geotextile. Many soil retention criterion are built around the particle size of the soil to be retained vis-a-vis the geotextile's opening size [3,4,5] and the general formulation is:

$$O_{95} \leq \lambda \cdot d_{85} \qquad (1)$$

where

O_{95} = 95% opening size, i.e., corresponding to the apparent opening size (AOS) of the geotextile,

d_{85} = 85% finer soil to be retained and

λ = a constant which varies from 1 to 5 in different formulations.

An accurate knowledge of an apparent opening size (AOS) is critical for predicting the filter behavior of the geotextiles. Rollin [3] pointed out that the existing ASTM method for determining AOS values does not lead to reproducible results. Lombard and Rollin [4] studied the significance of the opening size determination using various techniques such as dry sieving (AOS), wet sieving (FOS) and theoretically calculated values (COS) and found a significant difference among these values. It is believed that wet sieving (FOS)[5] results in a reliable measurement of an opening size of non-woven geotextiles.

The second and third characteristics; filtration ability and resistance to clogging, causes more difficulties in design. Recent research summarized by Carroll [6] has shown that the geotextile's opening size and permeability do not indicate clogging potential. Clogging potential can be evaluated in the laboratory using the gradient ratio test (GR). [2,6,7]. For granular soils and woven geotextiles (which were the focus during the initial geotextile development), the GR test is seen as a quite acceptable test to evaluate the clogging potential. However, for soils with significant amount of silt or clay fines or for gap-graded soils, the GR test is not acceptable [8]. For these types of soils, long-term filtration tests should be performed using the soil, the leachate and candidate geotextile for estimating the terminal flow rates and clogging potential [9]. At present, very little is known about the clogging mechanism especially for thick non-woven geotextiles and silty and gap-graded soils. Therefore, it is imperative to study the filtration behavior of non-woven geotextiles with silts and gap-graded soils emphasizing clogging problems of the soil-geotextile interface to develop design criterion.

In this study, long-term filtration tests are being conducted to evaluate the design life of non-woven geotextiles and to investigate clogging mechanisms which substantially reduce the design life of these geotextiles. In this study, bacterial clogging is not being considered, since there have been two major studies at L'ecole Polytechnique [10] and Drexel University [1] concentrating on this subject.

Materials, Testing Apparatus and Methodology

Two types of non-woven needle-punched geotextiles were used in this research. The geotextiles selected for this study have weight ranging from 133.3 to 450.0 gm/m^2 and their thickness ranged from 1.01 to 4.42 mm. Some important physical and hydraulic properties of these geotextiles are given in Table 1. The AOS values of these geotextiles were determined in the laboratory by dry sieving using glass beads from the Cataphote company. The filtration opening size (FOS) was determined at L'ecole Polytechnique by the wet sieving method using the glass beads. It can be seen that a significant difference exist between the AOS and FOS values for the given geotextiles. For the needle-punched fabric the measured FOS values decreased with increasing mass per unit area. This is consistent with the observation by Rigo et al. [11]. However, there seems to be no unique trend between FOS and mass per unit area for the heat-treated fabrics.

TABLE 1 -- Properties of the Needle-punched Geotextiles Tested.

Fabric Property	Non-treated			Heat-treated(one side)		
	1114	1125	1145	4NP	8NP	12NP
Weight, gm/m^2	140.0	246.6	450.0	133.3	266.6	400.0
Thickness, mm	1.64	2.78	4.42	1.01	2.02	3.0
Permittivity, sec.$^{-1}$	2.71	2.04	1.22	1.97	1.48	0.98
AOS(manfr's values, mm)	0.21-0.149	0.21-0.149	0.149-0.105	0.25-0.149	0.21-0.149	0.149-0.074
AOS(measured in lab, mm)	0.210	0.210	0.149	0.180	0.210	0.210
FOS (L'ecole Polytech., mm)	0.195	0.103	0.080	0.105	0.100	0.117

In Figure 1(a,b), the vertical sectional views of the untreated and heat-treated non-woven geotextiles are shown. As it can be seen from this figure, the untreated, needle-punched geotextile has a random structure and more pore space while the heat treated geotextile's fibers are oriented in straight weave and have less pore spaces.

Fig. 1a -- Electron micrograph of untreated needle-punched geotextile.

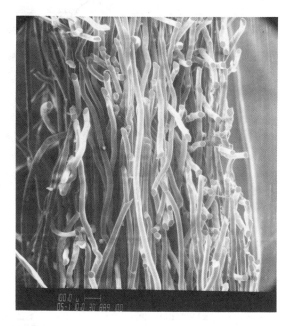

Fig. 1b -- Electron micrograph of heat-treated needle-punched geotextile.

Long-term filtration tests were performed using Ottawa sand and mixture of silt sized particles of sil-co-sil 75, 106 and 250 with Ottawa sand. Sil-co-sil product is produced from high purity silica that is ground to microcrystalline. The grain size distribution curves for all the mixtures are shown in Figure 2(a,b,c). The gap-graded nature of the soil tested is evident from Figure 2.

To conduct our long-term filtration tests, eight constant head permeameters were constructed. The permeameter consists of two plexiglass tubes, a smaller inner cylinder (diameter 9.5 cm) and a larger tube (diameter 20.3 cm) on the outside. All the tests were conducted for 100 to 200 hours at the hydraulic gradient of 10 and the height of the soil samples were 3.5 cm. In many applications of geotextiles as a filter, a lower value of hydraulic gradient is expected. However, to evaluate the worst clogging potential, a high value of the hydraulic gradient was selected. A total of 240 grams of soil was mixed with the same volume of deaired water (DOC values 5.38 ppm) to form a slurry. Various other sample preparation techniques were attempted and it was concluded that the slurry method produced a homogeneous mixture of sand and silts. Soil slurry was placed on top of the geotextiles in the permeameter and deaired and chlorinated water was introduced into the permeameter by showering as not to displace soil particles. After filling the inner cylinder with the water, the outer cylinder was filled. The first reading was taken fifteen minutes after the apparatus was initially filled. The tests were continued until no change in permeability was observed. After the test, the soil and geotextile samples were extruded and impregnated with ultra low viscosity epoxy. The encapsulated samples were cut and polished to be viewed through a reflected light microscope to study the microstructure of the clogged geotextile.

Retention tests on all geotextiles were performed. The filtered water (with silts) was put through the centrifuge to separate the water and soil particles. The soil was dried and weighed to determine the amount of soil passing through the geotextiles.

Test Results and Discussion

Series of tests were performed using all six geotextiles and eighteen mixtures of soils. Due to the length restrictions of this paper, only two series of tests results are discussed in detail. The results presented herein include those for the long-term filtration tests performed on a heat-treated (8NP) and untreated (1125) geotextile with mixtures of various percentages of sil-co-sil 75 and Ottawa sand.

From Figure 3 (Fabric 1125) the trend of the filtration curves is shown for the different percentages of soil mixed with the Ottawa sand. The first test in this series used sand without the addition of silt. The initial permeability of the system is 0.003 cm/sec, and stabilizes after approximately eighty hours at a value of 0.000145 cm/sec and continues to remain steady even after 100 hours. The shape of the curve indicates that the geotextile acts like an excellent filter, as the particles migrating towards it are stopped yielding a denser soil specimen and resulting in a cake formation upstream of the geotextile. The flow of water decreases with time and after a long filtration period, the permeability of the system reaches a constant value.

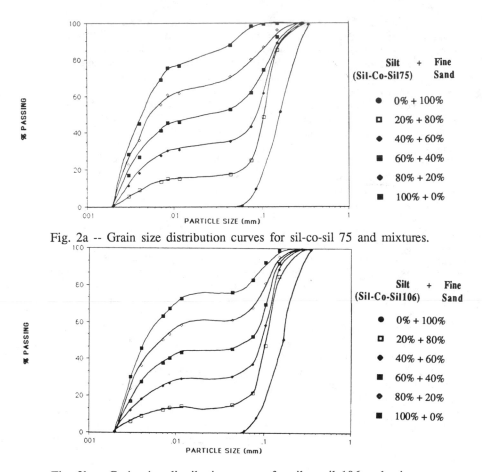

Fig. 2a -- Grain size distribution curves for sil-co-sil 75 and mixtures.

Fig. 2b -- Grain size distribution curves for sil-co-sil 106 and mixture

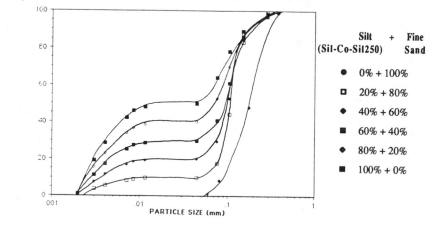

Fig. 2c -- Grain size distribution on curves for sil-co-sil 250 and mixtures.

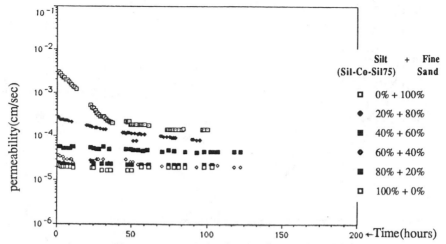

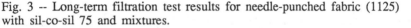

Fig. 3 -- Long-term filtration test results for needle-punched fabric (1125)
with sil-co-sil 75 and mixtures.

The addition of 20% of sil-co-sil 75 to the sand decreases the
permeability of the sample to an initial value of 0.000290 cm/sec as
compared to 0.003 cm/sec for the sand. Moreover, introducing fine
particles into the sands resulted in a closely packed or denser specimen.
When only sand is used, the uniform diameter of the sand particles form
large voids and the water can percolate through these voids at a higher
velocity. Sil-co-sil 75 particles, when mixed with the fine sand, are small
enough to interpenetrate the large voids filling the voids, consequently
decreasing the permeability of the system.

The filtration curve for the 40% silt test is lower than the previous
two tests. The addition of silt particles form a tighter structure creating a
flat shape for the long-term filtration curve which indicates faster cake
formation than the previous tests.

The filtration curve for the 60% silt with Ottawa sand has an even
lower value of permeability. During the initial 6 hours of the test a slight
decrease in permeability is seen. This is a result of the packing of the soil
particles and the formation of a filter cake. Other small fluctuations in
permeability may also be noticed and could be a result of rearranging
particles within the geotextile and soil structure.

The 80 and 100% curves are similar in both shape and magnitude.
Due to the fact that most of the particles are silty soil and the soil structure
is closely packed resulting in a very dense specimen. Verified by the final
permeability values of 60, 80, and 100% silts in all three tests, the same
final permeability value converges. The initial values of permeability are
higher for the 100% silt sample than for the 80% because initially the silt
particles float above the soil specimen in the inner cylinder. Particles
silting results in a smaller specimen height and higher permeability.

Figure 4 shows the reproducibility of the test results for both the 0%
and 100% sil-co-sil 75 tests. The permeability values are approximately the
same and the shape of the curves do not change dramatically. There are
some differences in the results mainly due to physical disturbances while
placing the soil specimen.

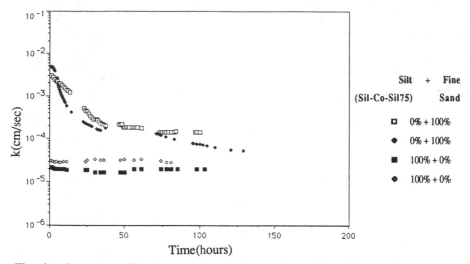

Fig. 4 -- Long-term filtration test results for needle-punched fabric (1125) with Ottawa sand and 100% sil-co-sil 75.

Figure 5 shows long-term filtration test results conducted on heat-treated geotextile using various percentages of sil-co-sil 75 mixed with fine sand. The results for the Ottawa sand is slightly different from the results of the untreated, needle-punched geotextile (see Figure 3). In this test, the permeability of the system initially decreases and after some duration of the test, it starts to increase again with reorientation of particles. The same test, using Ottawa sand, was conducted again to see the reproducibility of the results. From the trend of this second curve the same filtration behavior repeated, the curve initially decreased, then increased and later stabilized to approach a zero slope curve.

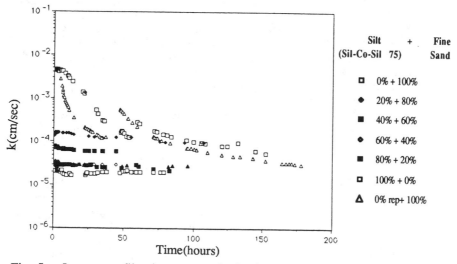

Fig. 5 -- Long-term filtration test results for heat-treated needle-punched fabric (8NP) with sil-co-sil 75 and mixtures.

The trends of the other curves are very much the same as discussed earlier for the non-woven needle-punched fabric. The 20% silt test performed on the heat-treated geotextile, however, showed the same fluctuations in behavior as for the 0% silt test.

The amount of soil passing the geotextiles were collected and measured to investigate the retention behavior of these geotextiles. Figure 6 shows the results in which the amount of soil passing is plotted vs. FOS/d_{85} of the soil. In this figure, only the most representative results are plotted to keep the clarity of the figure. The results show that the maximum amount of soil passing was for the thin needle-punched geotextile. The soil passing through the heat-treated fabrics were lower as compared to the untreated, needle-punched fabrics. These results were expected because the untreated, needle-punched fabrics show a larger mean distance between fiber, therefore, the larger pore channels allow more soil particles to pass through.

Our test results yielded a critical ratio of the filtration opening size (FOS) of the geotextiles to d_{85} of the soil being 3.0. This means that the ratio of FOS/d_{85} must be lower than 3 to retain soil particles and to form a stable system. It is important to point out that non-woven heat-treated fabrics had smaller amounts of soil passing. No relationship was found by using AOS values of the geotextiles.

The geotextiles used in the study by Faure et al. [12]) are similar to the ones used in the present research; however, the critical values of FOS/d_{85} were 2.0. One possibility for such a low critical FOS/d_{85} ratio could be because the soil used by Faure et al. are all poorly graded soils while the ones used in the present research are gap-graded soils. Gap-

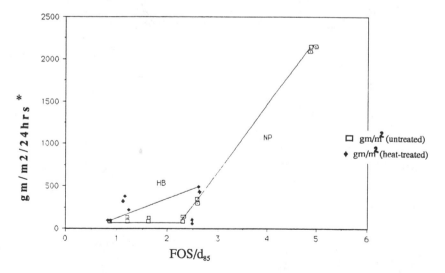

Fig. 6 -- Mass of soil passing through the geotextile during 24 hours versus FOS/d_{85}.

* wt. of the soil passing through the geotextile area in 24 hours.

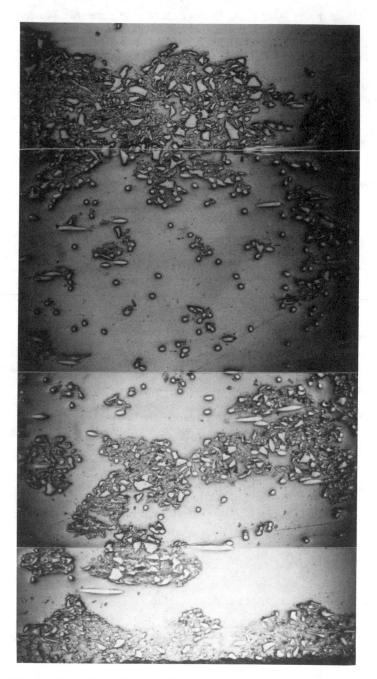

Fig. 7 -- Partially clogged view of needle-punched (1114) with sil-co-sil 75 (100%)(50x).

graded soils have a self-filtration mechanism which is not found in poorly graded soils. The self-filtration process operates as follows: the retained coarser particles filter fine particles that in turn filter finer ones. This process takes place until no more particles can migrate [13]. While this mechanism of gap-graded soils prevents excess particles from flowing through the geotextile, the poorly graded soils have no such mechanism and consequently loose more soil resulting in a lower critical FOS/d_{85} ratio.

Micrographs

Two series of micrographs for untreated, needle-punched and heat-treated fabrics are shown in Figures 7 and 8. The migration of fine particles in the geotextile structure result in an increasing clogged geotextile. It was viewed from the micrographs that with increasing percentages of fines, resulted in more clogged geotextiles. As can be viewed in Figure 7, significant amount of fine particles are embedded in the entire thickness of the geotextile. This clogging of geotextile takes place within first two to three hours of the test. During the same time, the soil particles are also settling and consolidating, therefore, it is difficult to separate the reduction in permeability due to clogging alone. It is important to note that none of the test resulted in a completely clogged geotextile. There were always some open pore space through which water could easily pass. Most of the clogging took place where the fibers were bunched together to form a pocket. One important conclusion that results from comparing micrographs of non-woven geotextiles of various thicknesses, is that the clogged structure of the geotextiles are very similar and thickness does not seem to effect the mechanism of clogging.

In Figure 8, a micrograph of a heat-treated geotextile is shown. This micrograph shows that the entire structure of the geotextile is clogged and only very few pores remain open for the water to flow. This clogged structure is very different as compared to the one shown in Figure 7 for the untreated, needle-punched geotextile. These results are expected because the untreated, needle-punched fabric has a large mean distance between the fibers as compared to the heat-treated; therefore, large channels in needle-punched geotextiles allow more soil to pass through the geotextile, resulting in a less clogged structure. This is also the reason that the thin untreated, needle-punched geotextile may have poorer soil retention performance than the thin heat-treated geotextile.

CONCLUSION

From the filtration tests and micrographs, four main points can be made regarding the filtration properties of non-woven geotextiles with gap-graded soils.

Gap-graded soils have a self filtration mechanism which enhances the filtration mechanism but also induces clogging in the geotextiles. As fine soil particles are added to the Ottawa sand the trend of the filtration curve becomes flatter indicating that there is not much change in the initial and final permeabilities of the soil-geotextile system. Due to the increasing gap-graded nature of the soil, the larger voids fill in with the smaller silt particles forming a denser specimen (or a tightly packed specimen).

One other important point to mention regarding the trend of the

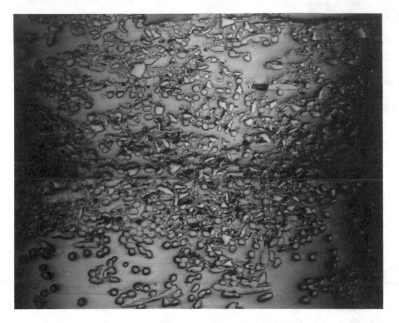

Fig. 8 -- Partially clogged view of heat-treated non-woven geotextile (4NP) with sil-co-sil 75 (100%) (50X).

filtration curves for both geotextiles (untreated and heat-treated, non-woven geotextiles) is that all curves follow a similar pattern. The heat-treated geotextile, however with coarser grained soils (with fines less than 30%) has an initial fluctuation in permeability. A possible explanation from Giroud's work [14]: The heat-treated, non-woven geotextiles have smooth finished surfaces allowing movements of coarse particles during the test due to the force of water on the specimen. The untreated, needle-punched geotextiles on the other hand are softer and rougher at the soil-geotextile interface preventing the above mentioned movement.

The retention criteria discussed in this paper shows a critical value of $FOS/d_{85} = 3.0$. This ratio is higher than the ones usually found in the literature due to the gap-graded nature of the soil. As mentioned earlier the gap-graded soils have a self filtration mechanism which maybe the cause of this higher FOS/d_{85} ratio. The self filtration process operates as follows: The retained coarser particles fill fine particles that in turn filter finer ones. This process continues until no more particles can migrate.

One conclusion shown by the micrographs is that as the fine content of the soil specimen increases, the depth of embedment of fines in the geotextile structure increases also. Particles are trapped in different quantities in the heat-treated and untreated, needle-punched geotextiles. Research is being conducted at Syracuse University to quantify the clogging in both types of non-woven geotextiles using image analysis methods.

Overall, there is not a significant difference in the filtration behaviors of untreated and treated, non-woven geotextiles. Both types of geotextiles get clogged but their long-term filtration performance was adequate.

ACKNOWLEDGEMENTS

The authors are very appreciative of the guidance provided by Professor R.M. Koerner of Drexel University, Professor A. Rollin and Y. Mlynarek of L'ecole Polytechnique. The support for this research was provided by the National Science Foundation and is gratefully acknowledged. Sincere appreciation is extended to Miss Heathere Booth for her careful typing of this manuscript.

REFERENCES

[1]   Koerner, G.R. and Koerner, R.M., "Biological Clogging in Leachate Collection Systems," Proceedings of a Seminar on Durability and Aging of Geosynthetics, Geosynthetic Research Institute, Drexel University, Dec. 8-9, 1988.

[2]   Calhoun, C.C., Jr., "Development of Design Criteria and Acceptance Specifications for Plastic Filter Cloth," Technical Report F-72-7, U.S. Army Corps of Engineers Waterway Experiment Station, Vicksburg, MS, 1972.

[3]   Rollin, A., "Filtration Opening Size of Geotextiles," ASTM Standard News, May, 1986, pp. 50-52.

[4]   Lombard, G. and Rollin, A., "Filtration Behavior Analysis of Thin Heat-treated Geotextiles," Proceedings of the Geosynthetics 1987 Conference, New Orleans, Session 4B, 1987, pp. 482-492.

[5]   Rollin, A.L., "CGSB Filtration Opening Size: Geotextiles Selection," Proceedings of the Third Canadian Symposium on Geosynthetics, Kitchener, Ontario, 1988, pp. 1-7.

[6]   Carroll, R.G., "Geotextile Filter Criteria," TRR 916, Engineering Fabrics in Transportation Construction, Washington, DC, 1983, pp. 46-53.

[7]   Haliburton, J.A. and Wood, P.D., "Evaluation of the U.S. Army Corps of Engineers Gradient Ration Test for Geotextile Performance." Proceedings Second International Conference on Geotextiles, Las Vegas, NV, Vol. I, 1982, pp. 97-101.

[8]   Koerner, R.M., Personal Communication, 1989.

[9]   Koerner, R.M. and Ko, F.K., "Laboratory Studies on Long-Term Drainage Capability of Geotextiles," Proceedings of the Second International Conference on Geotextiles, Las Vegas, NV, Vol. I, 1982, pp. 91-95.

[10]   Rollin, A., Andre, L. and Lombard, G., "Mechanisms Affecting Long-Term Filtration Behavior of Geotextiles," Journal of Geotextiles and Geomembranes, Vol. 7, 1988, pp. 119-145.

[11]   Rigo, J.M., Lhote, F., Rollin, A., Mlynarek, J. and Lombard, G., "Influence of the Geotextile Structure on Pore Size Determination," ASTM Microstructure of Geosynthetics Symposium, Orlando, FL, Jan. 1989.

[12]   Faure, Y., Giroud, J.P., Brochier, P. and Rollin, A., "Soil-Geotextile Interaction in Filter Systems," Proceedings of the Third International Conference on Geotextiles, Vienna, 1986, pp. 1207-1212.

[13]   Lafleur, J., Mlynarek, J. and Rollin, A.L., "Filtration of Broadly Graded Cohesionless Soils," Journal of Geotechnical Engineering, ASCE (to appear).

[14]   Giroud, J.P., "Filter Criteria for Geotextiles," Proceedings of the Second International Conference on Geotextiles, Las Vegas, NV, Vol. I, 1982, pp. 103-108.

John R. Rohde and Molly M. Gribb

BIOLOGICAL AND PARTICULATE CLOGGING OF GEOTEXTILE/SOIL FILTER SYSTEMS

REFERENCE: Rohde, J. R., and Gribb, M. M., "Biological and
Particulate Clogging of Geotextile/Soil Filter Systems",
Geosynthetic Testing for Waste Containment Applications, ASTM
STP 1081, Robert M. Koerner, editor, American Society for
Testing and Materials, Philadelphia, 1990.

ABSTRACT: Despite the increasing use of geotextiles in leachate
collection systems, the long-term filtration performance of
these materials has yet to be determined. A laboratory testing
program was undertaken to investigate biological and particulate
clogging of a typical geotextile. Four permeameters modeling a
leachate collection system were subjected to flow of municipal
landfill leachate under anaerobic conditions. Poisons were
utilized in half of the permeameters to inhibit biological
activity, thus allowing comparisons to be made between
biologically active and inactive systems. Volumetric flow rates
decreased 92 to 97% in all permeameters despite increases in
applied head from 6.5 to 35.5 cm. Large head losses were
observed in the first several centimeters of soil immediately
after the flow of leachate was initiated. Head losses over the
geotextiles increased as the study progressed, with the
unpoisoned geotextiles showing greater clogging.

KEYWORDS: geotextile, filtration, biological clogging, leachate
collection system, landfill design

In the past decade, geotextiles have grown in importance as
drainage and filtration media in civil engineering works. A recent
application of geotextiles is in primary leachate collection systems
of double liner sanitary landfills [1-5] (see Fig. 1). An EPA draft
[6] states that synthetic filter and drainage layers may be used in
leachate collection systems, providing that they meet the same
performance requirements as granular materials. A geotextile, when
included as a filter in a leachate collection system, must prevent
clogging of the drainage layer by waste materials and/or soil
particles. The permeability of the geotextile must remain greater

J. R. Rohde is an assistant professor and M. M. Gribb is a
research assistant at the University of Wisconsin-Milwaukee,
Department of Civil Engineering and Mechanics, Milwaukee, WI 53201.

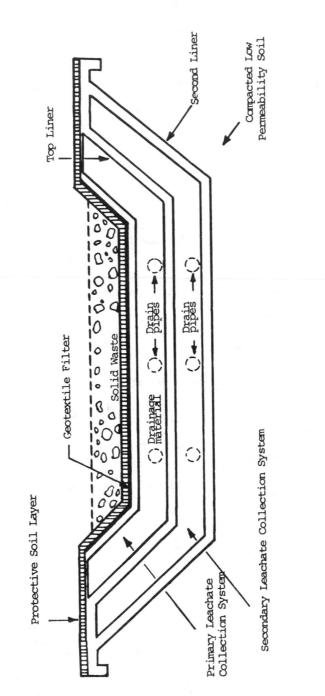

FIG. 1 -- Cross-sectional view of double liner landfill design containing primary and secondary leachate collection systems [6].

than or equal to the EPA designated permeability for a granular drainage layer, 0.01 cm/sec. To fully define the clogging phenomenon both biological and particulate processes must be considered.

The hydraulic behavior of geotextiles placed in contact with soil to facilitate the removal of water and/or prevent the movement of soil particles has been studied by a number of investigators [7-13]. However, few reports regarding the biological influences on clogging of geotextiles and soil/geotextile systems have been published. Ionescu et al. [14] incubated geotextiles in various media, including distilled water, sea water, compost, soil, and culture media for three types of bacteria for many as 17 months and found no significant loss of filtration efficiency. In contrast, Chen et al. [15] found significant reductions in the hydraulic capabilities of soil-geotextile filters due to microbial growth in the soil and in the geotextile itself.

Even less is known about the effects of municipal landfill leachate flow on geotextile filters. The presence of large numbers of microorganisms in landfill sites is well-documented. Fresh leachates have been characterized by very high biological oxygen demand (BOD) and chemical oxygen demand (COD) concentrations, 1 to 5 x 10^4 mg/L or higher [16]. Cancelli and Cazzuffi [17] exposed nonwoven geotextiles to leachate flow under aerobic conditions over a period of three hours and observed a substantial decrease in permeability normal to the plane of the fabric. This was attributed to the deposition of suspended solids on the surface of the geotextile and within the geotextile pore structure. Koerner and Koerner [18] immersed geotextiles in municipal landfill leachate under anaerobic conditions for 4 to 11 months. Once removed, the geotextiles displayed reductions in flow capacity of 5 to 20 percent. Koerner and Koerner determined that sand/geotextile systems subjected to aerobic flow of leachate for 4 to 11 months demonstrated decreased flow rates of 12 to 100 percent.

To date, the behavior of a geotextile filter subjected to the flow of leachate under anaerobic conditions as would be expected in the field has not been determined.

EXPERIMENTAL PROGRAM

A laboratory study was initiated to provide an assessment of biological and particulate clogging of a geotextile under landfill conditions. Four permeameters modeling a leachate collection system consisting of soil, geotextile filter, and drainage aggregate were constructed for this investigation. Well-graded sand was used for the soil component and a single polypropylene nonwoven geotextile was selected as the filter. The test columns were subjected to leachate flow under anaerobic conditions. Initially a constant head of 6.5 cm was applied over the system. As flow rates decreased, the head was increased to a maximum of 35.5 cm.

It is hypothesized that bacteria contribute to the clogging of filter systems through metabolic activities leading to the deposition of layers of biofouling material. Since this cannot be the case in a system free of viable microorganisms, it may be possible to draw inferences about the role of bacteria in clogging of geotextiles by comparing biologically active and inactive systems. To achieve this a biological poisoning scheme, similar to that described by Rosson, et al. [19], was incorporated into the experimental design. A wide spectrum inhibitor, mercuric chloride, was chosen for addition to one of the columns. In preliminary work, a layer of dark precipitates in the soil formed and progressed further down the soil column as the study continued. When these deposits were exposed to dilute HCl, vapors with a sulfuric odor evolved, suggesting that the precipitates were ferrous sulfide salts and possibly the result of microbial sulfate reduction. To determine if this was the case a specific inhibitor for sulfate-reducing bacteria, sodium molybdate, was selected for addition to a second column. The remaining two permeameter columns were not altered.

EXPERIMENTAL APPARATUS AND METHODS

Experimental Apparatus

Permeameter construction: The permeameters utilized in this study were made of clear cast acrylic pipe measuring 8.26 cm in diameter (i.d.). Four identical permeameters (Fig. 2) were constructed to accommodate a 35 cm soil layer and a 20 cm drainage aggregate layer separated by a geotextile filter. The permeameters consisted of two flanged sections to allow for the secure placement of the geotextile filters. An inlet port, 1.27 cm (i.d.), above the soil column allowed for introduction of leachate to the permeameter and an outlet port, of the same size, at the base of the aggregate permitted volumetric flow measurements to be taken. Piezometric ports placed at various levels along the upper section of the permeameter and 0.64 cm above the geotextile allowed for assessment of head loss along the soil column and differentiation between head losses in the soil column and over the geotextile. Swagelock-type fittings holding syringe septa at the top of the cells and beneath the geotextile surface permitted bacterial sampling and addition of poisons. The flanged sections were coated with vacuum grease and bolted together following column packing.

Leachate source: The leachate was obtained directly from an area municipal sanitary landfill about every three weeks during the study. The sample was drawn from a pipeline leading from the landfill's underground leachate collection system to a holding tank. Leachate samples were routinely assessed for bacterial content. Bacterial counts remained at the same order of magnitude, suggesting that the leachate could withstand storage in the laboratory and remain biologically active.

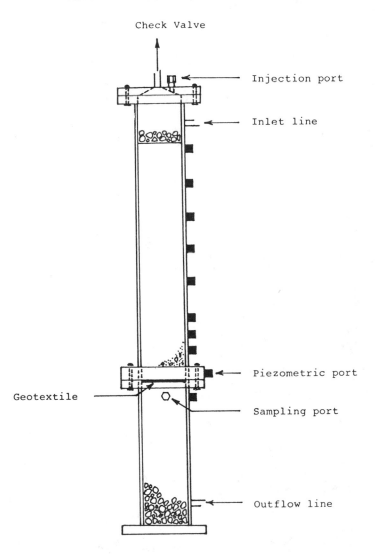

FIG. 2 -- Permeameter design.

Soil column components: A well-graded sand possessing a tap water permeability on the order of 2 X 10^{-2} cm/sec was selected. A grain size distribution of the test soil (Fig. 3) was determined by dry sieving according to ASTM Standard Method D-422. The sand was added to the columns in five to eight centimeter lifts, each being tamped twenty times with a steel rod in order to more uniformly compact the soil in the permeameters. A 20 cm layer of 1.0 cm-plus stream-run gravel was placed in the lower section of each permeameter column as the drainage aggregate component of the system. About five centimeters of gravel was placed above each soil column to retard the loss of soil during the initial backfilling stage of the test.

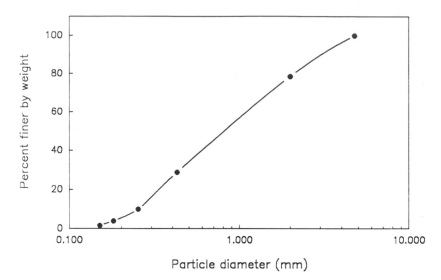

FIG. 3 -- Grain size distribution.

Geotextile properties: Polyfelt grade TS700, a nonwoven
polypropylene geotextile was utilized in this study. This geotextile
possesses an equivalent opening size (EOS) of 0.125 - 0.180 mm and a
permeability to water normal to the plane of the fabric of 0.5 cm/sec
(under 2.07 kPa of confining pressure). The unit weight and thickness
are 277.95 gr/m^2 and 2.66 mm, respectively [20].

System configuration: As shown in Fig. 4, the leachate supply was
held in a 60 liter barrel equipped with fittings to provide flow to
the constant head tank. A float valve in the tank regulated the fluid

FIG. 4 -- System configuration.

level to maintain a constant head level. A branched tubing arrangement was used to convey leachate to the permeameters, and a similar configuration on the down side of the cells allowed for effluent flow to the graduated cylinders used for flow rate measurements. To verify that the system remained free of pressure differences due to generation of methane or other gases, the source barrel and the constant head tank were connected to a manometer. A check valve on top of each permeameter allowed for venting of any gases which may have been produced in the columns.

Experimental Procedures

Initially, deaired tap water was used to backflush the permeameters in order to displace as much occluded air as possible. After 48 hours the direction of flow was reversed and water was allowed to flow through the system for three days prior to the introduction of leachate. To assure the anaerobic nature of the sample and testing environment, the sealed constant head tank was flushed with nitrogen prior to introduction of leachate.

Flow measurements: Volumetric flow rates were determined in each permeameter column at various intervals during the twelve week test. In addition, the local head losses along both the soil column and over the geotextile were determined by measuring piezometric water levels during the entire test period.

Poisoning regime: Two compounds, mercuric chloride and sodium molybdate were used as biological inhibitors in this test. The mercuric chloride was injected in one column using a Gilson peristaltic pump to maintain an end concentration of 0.001 Molar solution in an effort to inhibit all biological activity in that column. The sodium molybdate was pumped to another column to maintain an end concentration of 0.05 Molar solution in order to specifically inhibit sulfate-reducing bacteria. The remaining two columns were not poisoned.

Bacteriological analysis: Enumeration of the bacterial population of the unpoisoned leachate and that of leachate poisoned by mercuric chloride was accomplished by a direct count of colonies. In this way the effectiveness of the mercuric chloride in controlling bacterial growth could be assessed. The effectiveness of the sodium molybdate in the inhibition of sulfate-reducing bacteria could not be readily assessed via enumeration of the total bacterial population, and as such, samples from this column were not plated. Leachate samples were taken with small nitrogen-purged syringes from the septa on the permeameters. One-tenth ml aliquots of a ten-fold dilution series were plated on nutrient-rich L agar plates (5 grams yeast extract, 10 grams tryptone, and 10 grams sodium chloride per liter of distilled water) and spread with glass beads in accordance with basic microbiological techniques [21]. The L broth was chosen for the plating medium in order to support a wide variety of organisms. All dilutions were plated in quadruplicate and evaluated for aerobic and anaerobic bacterial growth by direct count of colonies. Enumeration of the anaerobic bacteria in samples was achieved by spreading and incubating half of the plates within an anaerobic glove bag.

RESULTS

Hydraulic Responses

At the onset of the test the 6.5 cm head induced a tap water flow averaging 0.25 ml/sec through the permeameter columns, which corresponds to k = 0.03 cm/sec for the sand/geotextile system. Upon introduction of leachate to the system, flow rates dropped to an average of 0.015 ml/sec; k = 0.002 cm/sec. The head over the system was increased in a stepwise manner during the twelve week testing period, to maintain a volumetric flow rate of 0.01 ml/sec, ensuring a complete turnover of leachate in the columns every 24 hours. The test was terminated when the head required to maintain a flow rate of 0.01 ml/sec exceeded 35.5 cm.

No statistically significant differences in the flow rates were observed among the four permeameters. The dominant mechanism in the reduction of flow in both the poisoned and unpoisoned columns was the accumulation of particulates in the top of the sand layers. As shown in Fig. 5, after twelve weeks of flow over 90% of the 35.5 cm of applied head was lost in the first 7.0 cm of sand. Because of these large head losses in the sand, the relative effects of geotextile clogging on the hydraulic responses of the overall systems were minimal.

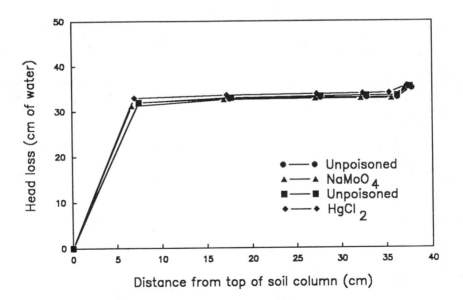

FIG. 5 --   Head loss versus depth for permeameter columns after 12 weeks of leachate flow. Geotextiles located 37.0 cm from top of soil columns.

Very little head loss occurred in the sand layers between a depth of 7.0 cm and the geotextiles as illustrated in Fig. 5. The head loss across the geotextile in the column poisoned by mercuric chloride (1.6 cm) was lower than that measured across the geotextiles in the unpoisoned columns (2.2 cm, 2.0 cm) or the column poisoned with sodium molybdate (2.1 cm).

The relationship between head loss and position in the permeameter for an unpoisoned column is shown in Fig. 6. The poisoned columns showed nearly identical relationships. While the rate of head loss decreased with time, the distribution through the columns remained consistent.

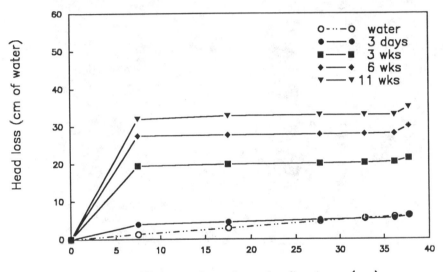

FIG. 6 -- Head loss versus depth in unpoisoned column.
         Geotextile located 37.0 cm from top of soil column.

Biological Responses

The appearance of the permeameter columns at the end of twelve weeks of flow is shown in Fig. 7. The addition of sodium molybdate to one of the columns resulted in an obvious reduction in the deposition of black solids, indicating that the presence of these solids was related to the presence of sulfate-reducing bacteria. A similar response was seen in the column poisoned with mercuric chloride. Furthermore, upon exposure to the atmosphere the dark deposits disappeared almost immediately. This suggested that these deposits were ferrous sulfide salts; which are known to develop under anaerobic conditions as a result of microbial sulfate reduction processes and are readily oxidized under atmospheric conditions. The mercuric chloride appeared to be effective in controlling bacterial growth in

the first column for the majority of testing period, as no viable
bacteria grew on agar plates incubated aerobically or anaerobically,
until the final plating. Plating after twelve weeks of flow indicated
a substantial number of viable bacteria in the mercuric chloride
poisoned column, suggesting an increasing population of mercury-
resistant bacteria. Aerobic and anaerobic bacteria in the influent
leachate were consistently on the order of 10^6/ml and 10^5/ml,
respectively, indicating that a relatively stable number of bacteria
were introduced to the system over time.

FIG. 7 --   Permeameter columns after 12 weeks of leachate flow.
             From left to right: Column 1, poisoned with mercuric
             chloride; column 2, unpoisoned; column 3, poisoned with
             sodium molybdate; column 4, unpoisoned.

To qualitatively examine the effects of biofouling on the
geotextile fiber surfaces, scanning electron microscopy (SEM) was
employed. Samples of a geotextile, extracted from an unpoisoned
column after twelve weeks, were prepared by fixation in glutaraldehyde
and by osmium tetroxide. After critical point drying, the samples
were sputter coated with gold. Fig. 8 shows a clean fiber, while Fig.
9 (a, b) shows a fiber exposed to leachate flow, and a higher
magnification of the exposed fiber, respectively. While Fig. 9 (a, b)
shows significant surficial bacterial fouling, observation of the
fabric skeleton did not reveal any macroscopic clogging. It should be
noted that the preparation of SEM samples required removal of all
detritus to prevent contamination of the microscope. As a result, it
was impossible to evaluate the amount of bacterial and particulate
matter held loosely within the fabric matrix prior to cleaning.

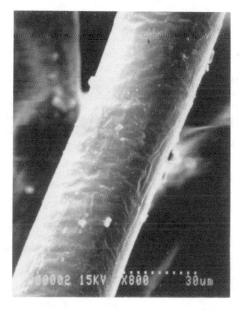

FIG. 8 -- A clean geotextile fiber at 800X.

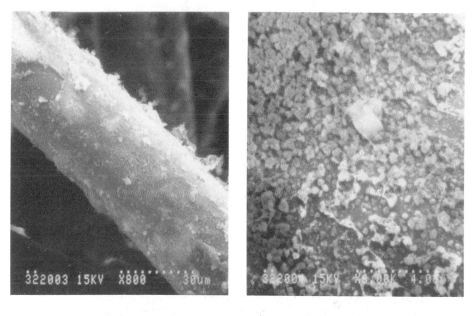

(a)   (b)

FIG. 9 -- A geotextile filter after 12 weeks of exposure to
          leachate flow, at (a) 800X and (b) 6000X.

CONCLUSIONS

Anaerobic conditions were maintained over an extended period of time in a permeameter system closely modeling a municipal landfill leachate collection system.

Poisons provide an effective means for the isolation of the effects of biofouling and particulate clogging. However, the complex variety of bacterial organisms in municipal landfill leachate makes complete inhibition difficult. Reactivity of the poisons with the leachate must be contemplated.

Biofouling may affect the filtration performance of geotextiles.

It is important to consider the entire soil/geotextile system when designing a filter.

There is a need for a long-term study to evaluate the adequacy of the present design standards.

ACKNOWLEDGEMENTS

This study was supported by the University of Wisconsin-Milwaukee Graduate School and Dr. Rohde's Presidential Young Investigators Grant from the National Science Foundation.

REFERENCES

[1] Lundell, C. M., and Menoff, S. D., "The Use of Geosynthetics As Drainage Media at Solid Waste Landfills", Proceedings of the Geosynthetic '89 Conference, Vol. 1, Industrial Fabrics Association International, St. Paul, 1989, pp. 10-17.

[2] Buranek, D., and Pacey, J., "Geomembrane-Soil Composite Lining Systems Design, Construction Problems, and Solutions", Proceedings of the Geosynthetic '87 Conference, Vol. 2, Industrial Fabrics Association International, St. Paul, 1987, pp. 375-384.

[3] Haxo, H. E., and Waller, M. J., "Laboratory Testing of Geosynthetics and Plastic Pipe for Double Liner Systems", Proceedings of the Third International Conference on Geotextiles, Vol. 2, Industrial Fabrics Association International, St. Paul, 1986, pp. 577-593.

[4] Rollin, A. L., and Denis, R., "Geosynthetic Filtration in Landfill Design", Proceedings of the Geosynthetic '87 Conference, Vol. 2, Industrial Fabrics Association International, St. Paul, 1987, pp. 456-470.

[5] Yamamoto, L. O. "Design and Construction of a Hazardous Waste Landfill", Proceedings of the Geosynthetic '87 Conference, Vol. 2, Industrial Fabrics Association International, St. Paul, 1987, pp. 353-364.

[6] Environmental Protection Agency, "Minimum Technology Guidance on Double Liner Systems for Landfills and Surface Impoundments--Design, Construction, and Operation", EPA/530-SW-85-014, Environmental Protection Agency, Washington, DC, 1985.

[7] Calhoun, C. C., "Development of Design Criteria and Acceptance Specifications for Plastics Filter Cloths", Technical Report S-72-7, United States Army Engineer Waterways Experimental Station, Vicksburg, 1972.

[8] Koerner, R. M., and Ko, F. K., "Laboratory Studies on Long-Term Drainage Capability of Geotextiles", Proceedings of the Second International Conference on Geotextiles, Vol. 1, Industrial Fabrics Association International, St. Paul, 1982, pp. 91-95.

[9] Marks, B. D., "The Behavior of Aggregate and Fabric Filters in Subdrainage Applications", University of Tennessee, Knoxville, 1975.

[10] Schober, W., and Teindl, H., "Filter Criterion for Geotextiles", Proceedings of the Seventh European Conference on Soil Mechanics and Foundation Engineering, Vol. 2, British Geotechnical Society, 1979, pp. 121-129.

[11] Scott, J. D., "The Filtration-Permeability Test", Proceedings of the First Canadian Symposium on Geotextiles, Canadian Geotechnical Society, Rexdale, Ontario, Sept. 1980, pp. 176-186.

[12] McKeand, E., "The Behavior of Nonwoven Fabric Filters in Subdrainage Applications", Proceedings of the International Conference on the Use of Fabrics in Geotechnics, Ecole Nationale des Ponts et Chaussees, Laboratoire Central des Ponts et Chaussees, Paris, 1977, pp. 171-176.

[13] Lombard, G. and Rollin, A. L., "Filtration Behavior Analysis of Thin Heat-Bonded Geotextiles", Proceedings of the Geosynthetic '87 Conference, Vol. 2, Industrial Fabrics Association International, St. Paul, 1987, pp. 482-492.

[14] Ionescu, A., Kiss, S., Dragan-Bularda, M., Radulescu, D., Kolozsi, M., Pintea, F., and Crisan, R., "Methods Used for Testing the Bio-Colmatation and Degradation of Geotextiles Manufactured in Romania", Proceedings of the Second International Conference on Geotextiles, Vol. 2, Industrial Fabrics Association International, St. Paul, 1982, pp. 547-552.

[15] Chen, Y. H., Simons, D. B., and Demery, P. M., "Hydraulic Testing of Plastic Filter Fabrics", *Journal of the Irrigation and Drainage Division, ASCE*, Vol. 107, No. IR3, 1981, pp. 307-325.

[16] Lu, J. C., Eichenberger, B., and Stearns, R. J., *Leachate from Municipal Landfills: Production and Management*, Noyes Publications, Park Ridge, 1985.

[17] Cancelli, A., and Cazzuffi, D., "Permittivity of Geotextiles in Presence of Water and Pollutant Fluids", *Proceedings of the Geosynthetic '87 Conference*, Vol. 2, Industrial Fabrics Association International, St. Paul, 1982, pp. 471-481.

[18] Koerner, G. R., and Koerner, R. M., "Biological Clogging in Leachate Collection Systems:, in *Proceedings of a Seminar on Durability and Aging of Geosynthetics*, Geosynthetic Research Institute, Philadelphia, 1988.

[19] Rosson, R. A., Tebo, B. M., and Nealson, K. H., "Use of Poisons in Determination of Microbial Manganese Binding Rates in Seawater", *Applied and Environmental Microbiology*, Vol. 47, No. 4, 1984, pp. 740-745.

[20] Polyfelt Inc., product specification sheet, Evergreen, Alabama.

[21] Gerhardt, P., Murray, R., Costilow, R., Nester, E., Wood, W., Krieg, N., and Phillips, G., Eds., *Manual of Methods for General Bacteriology*, American Society for Microbiology, Washington, DC, 1981.

George R. Koerner and Robert M. Koerner

BIOLOGICAL ACTIVITY AND POTENTIAL REMEDIATION INVOLVING
GEOTEXTILE LANDFILL LEACHATE FILTERS

REFERENCE: Koerner, G. R. and Koerner, R. M., "Biological
Activity and Potential Remediation Involving Geotextile
Landfill Leachate Filters," "Geosynthetic Testing for Waste
Containment Applications, ASTM STP 1081, R. M. Koerner, Ed.,
American Society for Testing and Materials, Philadelphia, 1990

ABSTRACT: This paper presents the results of a biological
growth study in geotextile filters used in landfill leachate
collection systems. After reviewing the first year's activity,
a completely new experimental approach has been taken. Using
100 mm diameter columns for the experimental incubation and
flow systems, the effects of six landfill leachates are
evaluated. Aerobic and anaerobic states, four different
geotextiles, and soil/no soil conditions above the geotextiles
are involved in the testing program. This results in 96
individual test columns. Flow data is measured regularly, and
over the first six months of evaluation the following trends
have been observed.
   • no clogging (0%-25% flow reduction)
      6 of 96 columns = 7%
   • minor clogging (25%-50% flow reduction)
      4 of 96 columns = 4%
   • moderate clogging (50%-75% flow reduction)
      37 of 96 columns = 38%
   • major clogging (75%-95% flow reduction)
      35 of 96 columns = 36%
   • severe clogging (95%-100% flow reduction)
      14 of 96 columns = 15%
For two of the landfill leachates, backflushing has been
attempted so as to reinstitute flow. This procedure works well
for the geotextile alone while not as well for the geotextile/
soil columns. The exception is the nonwoven heat set geotextile. All tests are still ongoing and will be dismantled and
further investigated at the end of 12 months exposure time. The
experimental setup and procedure has been written up as a ten-
tative ASTM test method and is currently in task group review.

   G. R. Koerner is Senior Research Specialist and R. M. Koerner is
Director and Professor at the Geosynthetic Research Institute, Drexel
University, Philadelphia, PA 19104

KEYWORDS: biological clogging, aerobic, anaerobic, soil clogging,
geotextile clogging, filtration, leachate, bacteria count, viable
count

INTRODUCTION

    The leachate collection systems used above and below primary
liners in landfills are meant to function in a free flowing
gravitational mode for their entire active and post closure care
periods. Such leachate collection systems consist of a drainage
material (either sand, gravel or a geonet), a protective filter
(either sand or a geotextile), and sometimes a perforated pipe covered
with an appropriate filter (either sand, gravel or a geotextile).
Figure 1 shows the location of these materials where geosynthetic
alternates are illustrated. It should be noted that a fine grained
soil (silt, clay or mixture) is frequently placed between the leachate
collection system and the waste. This layer if often referred to as
an "operations layer" or "working surface". Due to its small void
structure it could filter out micro-organisms and fine sediment in the
leachate. This condition is not modeled in the tests to be described
in this paper. While both natural materials and their geosynthetic
counterparts can be designed using state-of-the-art techniques and
test methods, the general focus is usually on short-term performance.
When periods of 30 to 50 years are required, long-term concerns must
also be addressed. Obviously, durability and aging concerns are very
important [1,2], but concern that the leachate collection system
remains free flowing is also important. With clogging will come a
buildup of hydraulic head on the filter, creating a zone of saturation
within the waste. Thus long-term clogging becomes an issue since it
can arise from either particulate or biological mechanisms.
Particulate clogging has been evaluated in a number of test simu-
lations, e.g. gradient ratio [3], long-term flow [4] and hydraulic
conductivity ratio [5] tests. Biological clogging, however, has only
recently been addressed [6] and/or evaluated [7,8]. This paper is
focused toward a continuation of an earlier report on biological
clogging of geotextiles and represents modifications and extensions of
that initial one year effort.

OVERVIEW OF FIRST PHASE OF PROJECT

    It is well recognized that municipal landfill leachates contain
large amounts of various microorganisms, primarily different forms of
bacteria. Figure 2 shows the biological counts at six landfills in the
northeast region of the USA. The total direct count measures all
bacteria, while the viable titer gives the living bacteria count in
units of number of cells per milliliter of leachate. Note the
magnitude of these numbers. Chemical analysis data on these same six
landfill leachates are given in Table 1. It is evident from this data
that each leachate is unique. The BOD_5 values are generally considered
to be the best indicators of the available biological activity.

    This information certainly suggests that biological activity is
present and when combined with moderate-to-warm temperatures (as
occurs at the bottom of a landfill) and an ample food source (as
contained in domestic waste), the growth of bacteria within the

TABLE 1 -- Details of municipal landfill leachates evaluated in this study and approximate leachate characteristics after first year's study.

Site Designation	Leachate Management Scheme	Approximate Leachate Characteristics at Project Startup			
		pH	COD (mg/l)	TS (mg/l)	BOD$_5$
PA-1	Continuously Removed	8.0	15,000	8,000	2,000
NY-2	Recycled through Landfill and Continuously Removed	5.5	12,000	7,000	3,000
DE-3	Recycled through Landfill	5.8	40,000	17,000	24,000
NJ-4	Continuously Removed	7.4	45,000	16,000	25,000
MD-5	Continuously Removed	6.8	5,000	2,000	1,000
PA-6	Continuously Removed	6.5	10,000	5,000	2,500

where

COD   =   chemical oxygen demand
TS   =   total solids content
BOD$_5$ =   biochemical oxygen demand at five days

leachate collection systems is certainly possible. Further
consideration of the situation would suggest that the filter (rather
than the drain) should be the focus of attention since it contains the
smallest voids through which the leachate must flow. This is the case
for both fine to medium sand filters and geotextile filters since
their void diameters are approximately the same, however, their
thicknesses are significantly different. The point to be made is that
both sand and geotextile filters should be evaluated for their
biological clogging potential.

The first phase of this project [9] evaluated seven different
geotextiles (a minimum of four per site) in aerobic flow boxes with
sand above them, and an additional four in anaerobic incubation drums
with subsequent flow and strength tests. The study was performed at
six landfill sites and lasted for twelve months. From the aerobic flow
tests it was found [9];

(a) that flow rate reductions were from 40% to 100%,
(b) that geotextile opening size played a key role, with larger sizes
    allowing for passage of the clogging sediment and/or dormant
    bacteria,
(c) that the type of geotextile polymer is of no great significance,
(d) that soil clogging could not be separated from geotextile
    clogging, and
(e) that particulate clogging could not be distinguished from
    biological clogging.

From the anaerobically incubated samples it was found [9];

(a) that flow rate reductions were from 10% to 40%,
(b) that the biological buildup was cumulative as confirmed by
    photomicrographs which showed progressively greater biological
    attachment over the 12 month testing period, see Figure 3,
(c) that there was no physical attachment of the biological growth to
    the geotextile fibers, and
(d) that there was no strength loss of the geotextile over the
    12-month incubation period

Building upon these results, a second phase of the project was
aimed at eliminating the objectionable features of the first phase and
providing for an opportunity to remediate the filtration systems by
backflushing. The results of this second phase activity follows for
the remainder of the paper.

DETAILS OF CURRENT PROJECT

It is felt that, the new test columns for this second phase of the
study of biological activity in landfill filters must meet the
following criteria.

(a) Sand filter clogging should be distinguishable from geotextile
    filter clogging.
(b) Particulate clogging should be distinguishable from biological
    clogging.
(c) Partly saturated (aerobic) clogging should be distinguishable from
    saturated (anaerobic) clogging.

(d) Identical geotextiles should be used at every site.
(e) The flow columns should be capable of accommodating continuous or periodic flow testing.
(f) The flow columns should use the leachate at the time of testing and not be stored for any length of time least it change in its composition.
(g) Constant head or variable head conditions should be capable of being accommodated.
(h) The flow columns should be capable of being backflushed with leachate and the results assessed.
(i) The flow columns should be capable of being flushed from either side with biocide and the results assessed.

In order to meet these needs, flow columns as shown in Figure 4 have been developed and are used in this second phase study. It must be cautioned, however, that some owners or agencies may not allow backflushing as a remediation method.

The flow columns of Figure 4 are constructed out of commonly available 100 mm diameter PVC pipe and related fittings. The containment ring is actually a pipe coupling which has a raised inner "lip" upon which the geotextile is placed and sealed. A non-water soluble adhesive is used to bond the geotextile to the lip and to prevent edge leakage. The upper and lower tubes are both 100 mm long pieces of pipe and they are contained by end caps which have 25 mm holes pre-drilled in them and are threaded. Support gravel is placed below the geotextile prior to positioning and gluing the lower end cap. Similarly, if soil is to be placed above the geotextile it must be done before the upper end cap fixed. End cap adaptors are then threaded into the end caps and fitted with 25 mm flexible tubing (for constant head tests) or rigid tubing (for variable head tests). These two options are shown in Figures 5 and 6, respectively, along with photographs of the completed devices. The experimental design for this second phase study was as follows:

• Four identical (continuous filament) geotextiles were used at each site and under each set of conditions:

  • 240 g/m² woven monofilament of 0.21 mm average opening size and 6% open area
  • 140 g/m² nonwoven heat set of 0.21 to 0.15 mm average opening size
  • 270 g/m² nonwoven needle punched of 0.21 mm average opening size
  • 540 g/m² nonwoven needle punched of 0.15 mm average opening size

• Soil (uniformly graded Ottawa sand of 0.42 mm average size) was placed above one set of the geotextiles, while nothing was placed above another set.
• One set of all of the above mentioned columns was allowed to drain between readings (thus providing aerobic conditions), while another set was constantly immersed in leachate (thus providing essentially anaerobic conditions). Note that throughout this paper we will refer to this setup as being anaerobic due to its full saturation conditions. It is very possible, and perhaps even likely, that some small amount of air enters the system greatly complicating the actual bacterial composition.
• All of the above variations were done at each of the six landfill

sites, thus 96 (4 × 2 × 2 × 6) flow columns of the type shown in
Figure 4 are included in this study.

RESULTS OF CURRENT PROJECT

This section describes the results of individual studies using the
flow columns just described. The subsections to be described are, (a)
continuous short term flow tests, (b) periodic long term flow results,
and (c) the effects of leachate backflushing.

(a) Short-Term Continuous Flow Tests

Since all of the tests during the first year were performed on a
monthly basis, and the distinction between fine particulate clogging
versus biological clogging was never settled, a set of continuous flow
tests were performed. Here the flow columns were set up in a variable
head mode, as shown in Figure 6, and leachate was continuously
supplied directly from a leachate sump and passed through the system.
The geotextile/soil configuration was used so that flow times were
long enough to be accurately measured. The results of this testing at
the two sites with the harshest leachates, DE-3 and NJ-4, are shown in
Figure 7. After an initial decrease which was probably a tuning of the
soil/geotextile system to the flow regime and the formation of a
stable flow network, the permeability of each leachate leveled off to
essentially constant values. Thus it was felt that what sediment is in
the leachate does not continue to build up so as to stop, or even
substantially decrease, the system's flow. This suggests that the
short term filtration characteristics of the soil and the geotextile
are adequate to handle the indicated flow rates. It furthermore,
provides a reference plane to which the long-term flow rates can be
compared. Such long-term flow tests are the focus of the next
section.

(b) Long-Term Intermittent Flow Tests

Long term flow evaluation of the columns at all six landfill sites
were undertaken. Variable head tests of the sixteen variations at each
site were performed for six months. Figures 8, 9, 10 and 11 give these
results for each of the four geotextiles mentioned in the previous
section. They are the woven, nonwoven heat-set, light nonwoven needled
and heavy nonwoven needled geotextiles respectively. The anaerobic
results are on the left sides of each figure and the aerobic results
are on the right sides. The soil covered geotextiles are the upper
curves, while the geotextiles by themselves are the lower curves for
each figure. The coding on the graphs for the various test conditions
is as follows.

WM(N)-PP   = woven monofilament (non-calendared) polypropylene
NW(HS)-PP   = nonwoven (heat set) polypropylene
NW(N)-PET 8 oz   = nonwoven (needled) polyester of 8 oz/yd^2 weight
NW(N)-PET 16 oz = nonwoven (needled) polyester of 16 oz/yd^2 weight

AN/S = anaerobic condition with sand above
A/S = aerobic condition with sand above
AN/W = anaerobic condition without sand above
A/S = aerobic condition with sand above

Some observations on the trends observed in Figures 8 to 11 are worthy of note.

- The anaerobic flow behavior is remarkably similar to the aerobic flow trends insofar as the system permeability is concerned.
- The tests with sand above the geotextiles are much smoother in their trends than those of the geotextiles alone which have very abrupt changes in permeability.
- In general, the sand/geotextile systems gradually decreased in their permeability with the nonwoven heat set geotextile of Figure 9 showing the greatest decrease after six months.
- Viewing the entire set of data collectively, we find the following:
  - no clogging (0%-25% flow reduction)
    6 of 96 columns = 7%
  - minor clogging (25%-50% flow reduction)
    4 of 96 columns = 4%
  - moderate clogging (50%-75% flow reduction)
    37 of 96 columns = 38%
  - major clogging (75%-95% flow reduction)
    35 of 96 columns = 36%
  - severe clogging (95%-100% flow reduction)
    14 of 96 columns = 15%

- Within this group, the leachates of DE-3 and NJ-4 resulted in the greatest amount of clogging. They will be focused upon in the next section.

(c) Leachate Backflushing Tests

Paralleling efforts in the sewer pipe cleaning, agricultural drain cleaning and sewage treatment filter cleaning businesses, it appears worthwhile that we should attempt backflushing. Each column was backflushed using the site specific leachate at the end of its six month incubation, i.e., at the terminus of the graphs shown in Figures 8 to 11. Backflushing was done from the bottom of the geotextile at a constant head of 60 cm for a period of 15 minutes. The head was sufficiently low so as not to have liquifaction of the sand above the geotextile for those cases where the system had sand. The percent recovery determined by performing regular flow tests after the backflushing is given in Table 2. Note that in all cases the geotextile by itself was restored to a higher recovery flow rate than the sand/geotextile combinations with the exception of the nonwoven heat set geotextile. When the columns are dismantled we will examine this situation carefully.

Within the following month after this flow rate recovery, flow again was seen to decrease. These trends, however, are still being developed. We anticipate patterns such as illustrated in Figure 12. A number of features of these curves are of significance. They show the periodicity of required backflushing, how biocide introduced into the backflush affects the situation, and the net recovery reinstated after each backflushing. Work is ongoing in this regard.

TABLE 2 -- Percent flow rate recovery after leachate backflushing at 60 cm head for 15 minutes.

Biological Condition	Material Above Geotextile	Geotextile Type	Landfill Site DE-3	NJ-4
anaerobic	sand	woven monofilament	60%	50%
aerobic	sand	woven monofilament	40	40
anaerobic	no sand	woven monofilament	100	100
aerobic	no sand	woven monofilament	100	100
anaerobic	sand	nonwoven heat set	60	45
aerobic	sand	nonwoven heat set	40	40
anaerobic	no sand	nonwoven heat set	5	10
aerobic	no sand	nonwoven heat set	5	5
anaerobic	sand	light nonwoven needled	60	60
aerobic	sand	light nonwoven needled	60	40
anaerobic	no sand	light nonwoven needled	80	75
aerobic	no sand	light nonwoven needled	60	60
anaerobic	sand	heavy nonwoven needled	60	55
aerobic	sand	heavy nonwoven needled	60	40
anaerobic	no sand	heavy nonwoven needled	100	85
aerobic	no sand	heavy nonwoven needled	50	65

SUMMARY AND CONCLUSIONS

The long term drainage of leachate collection systems at landfill sites is of major importance in understanding leachate management strategies. If the filters for such drains clog (via either particulate or biological activity), the hydraulic head on the filter will increase, forcing saturated leachate conditions into the waste mass itself. The indications from the first year's study of this project, and the repetition with a greatly improved containment device over a subsequent six months, strongly suggests that such clogging and leachate buildup will occur. From the second generation flow devices presented in this paper it appears as though the majority of clogging is biologically oriented rather than particulate. The times for severe clogging (arbitrarily defined as a flow reduction of 95% or more) for the different soil/geotextile and geotextile systems are relatively short. It was seen that the geotextiles by themselves exhibited a dramatic decrease in flow soon after biological activity initiated. In contrast, the geotextile/soil systems exhibited gradual decreases in flow after biological activity initiated. We feel that the soil affords a thickness (or buffering) effect which is not available to the geotextile by itself.

In order to alleviate the clogging, leachate backflushing tests were performed on all 96 incubation devices. The improvement was remarkable: approximately 51% flow rate increase for the sand/geotextile combinations and 63% for the geotextiles by themselves. Of course, some (or all) of the clogging may return. That is precisely the stage we are currently investigating.

In closing it should be emphasized that many different leachate collection cross sections are possible and we have examined only two of them, i.e., the geotextile by itself and geotextile with an overlying medium rounded sand layer. Different soil types, the existence of operation's layers or working surfaces, etc., will all influence the leachate flow regime and have different implications. Thus site-specific modeling of the intended cross section should be undertaken. This paper has given a experimental method and procedure to accomplish this type of modeling. Regarding the use of backflushing to relieve clogging of leachate filters it should be cautioned that the removal system must be designed accordingly and approval of this approach must be gained during the permitting process. It should also be recognized that the backflushing liquid, if water, will add to the leachate quantity to be eventually treated and is generally not desirable from an operations point of view. Thus air (or nitrogen) backflushing or vacuum withdrawal might be desirable options. Both of these possibilities are currently being evaluated.

ACKNOWLEDGEMENTS

This project is funded by the U.S. Environmental Protection Agency under Project No. CR-814965-02. Our sincere appreciation is extended to the Agency and in particular our Project Officer, Robert E. Landreth.

REFERENCES

[1] Durability and Long Term Performance of Geotextiles, "K. Gamski, Editor, Elsevier Applied Science Publ., Vol. 7, Nos. 1 and 2, 1988.
[2] Aging and Durability of Geosynthetics, R. M. Koerner, Editor, Elsevier Applied Science, 1989.
[3] Haliburton, T. A. and Wood, P. D., "Evaluation of U.S. Army Corps of Engineers Gradient Ratio Test for Geotextile Performance," Proceedings of the 2nd International Conference on Geotextiles, Las Vegas, NV, Aug. 1-6, 1982, IFAI, pp. 97-101.
[4] Halse, Y., Koerner, R. M. and Lord, A. E., "Filtration Properties of Geotextiles Under Long Term Testing," Proceedings of the ASCE/PennDOT Conference on Advances in Geotechnical Engineering, Hershey, PA, Apr. 1987, pp. 1-13.
[5] Williams, N. D. and Abruzakhm, M. A., "Evaluation of Geotextile/Soil Filtration Characteristics Using the Hydraulic Conductivity Ratio Analysis," Journal of Geotextile and Geomembranes, Elsevier Applied Science Publ., Vol. 8, No. 1, 1989, pp. 1-26.
[6] Ramke, H.-G., "Consideration on the Construction and Maintenance of Dewatering Systems for Domestic Trash Dumps," TR087-0119, translated from German by U.S. EPA, Cincinnati, Ohio, Nov. 1987.
[7] Rios, N. and Gealt, M. A., "Biological Growth in Landfill Leachate Collection Systems," in Durability and Aging of Geosynthetics, Elsevier Applied Science Publ., 1989, pp. 244-259.

[8] Koerner, G.R. and Koerner, R. M., "Biological Clogging in Leachate
     Collection Systems," in <u>Durability and Aging and Geosynthetics</u>,
     Elsevier Applied Science Publ., 1989, pp. 260-277.

[9] Koerner, G. R. and Koerner, R. M., "Biological Clogging of
     Geotextiles Used as Landfill Filters; First Year Results," GRI
     Report #3, Geosynthetic Research Institute, Philadelphia, PA, June
     27, 1989.

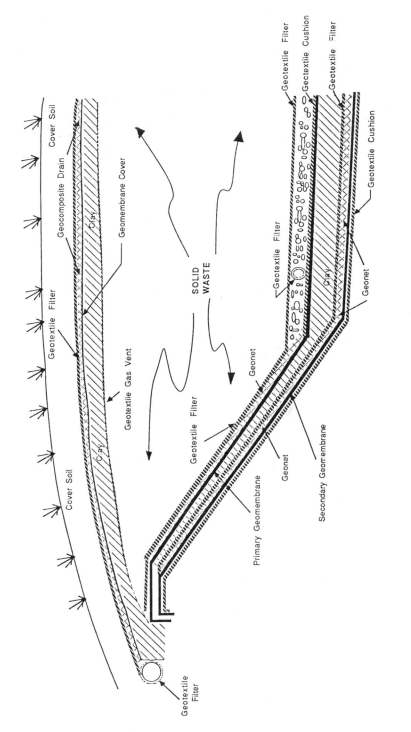

FIG. 1 -- "Typical" landfill liner and cover system.

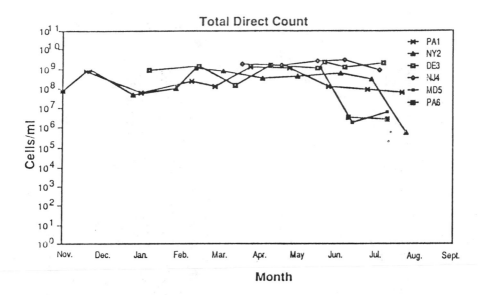

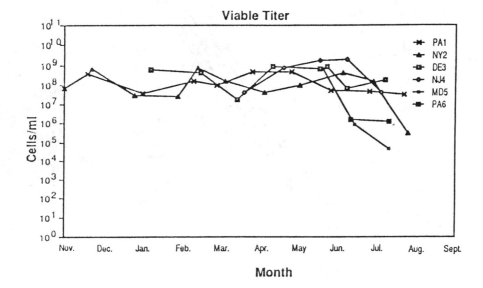

FIG. 2 -- Total bacteria count and viable (living) count of leachate
samples from the six landfill sites evaluated in this study,
after Rios and Gealt [7].

PA1 NW (N)-PP2 400X 6M   PA1 NW (N)-PP2 400X 12M

FIG. 3 -- Scanning electron micrographs of biological growth on
         geotextile fibers after 1, 3, 6 and 12 month anaerobic
         incubation.

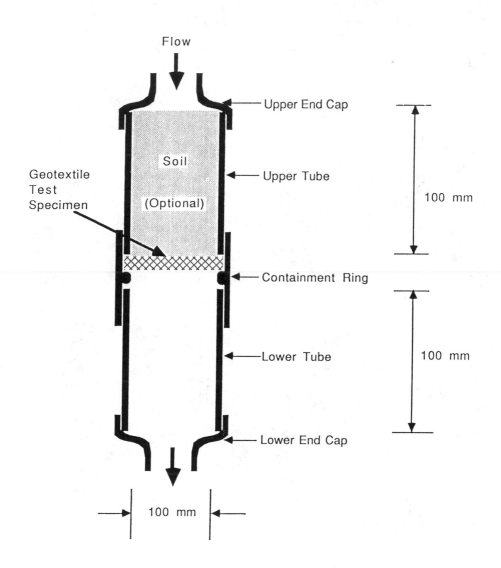

FIG. 4 -- Flow column used to contain geotextile test specimen and
optional use of soil/geotextile combined systems.

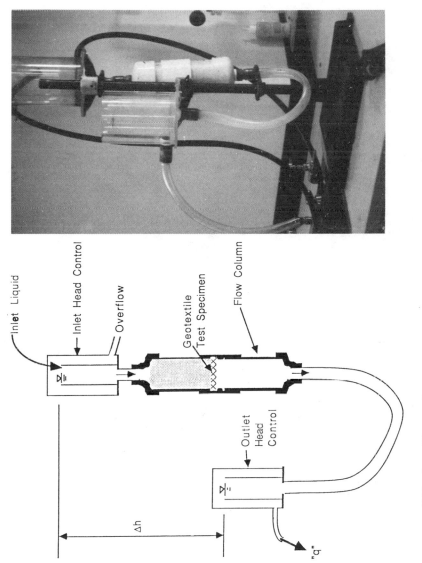

FIG. 5 -- Flow column and hydraulic head control devices for constant head tests.

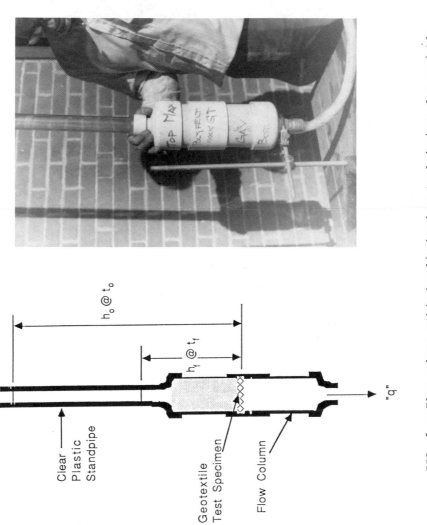

FIG. 6 -- Flow column and hydraulic head control devices for variable (or falling) head tests.

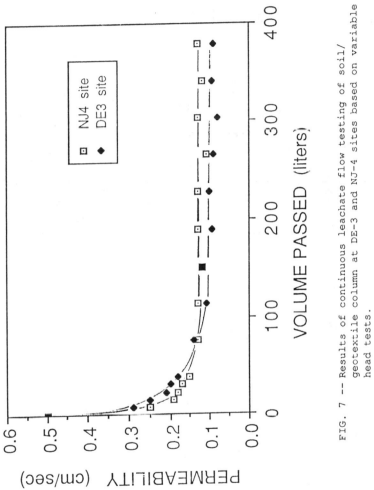

FIG. 7 -- Results of continuous leachate flow testing of soil/
geotextile column at DE-3 and NJ-4 sites based on variable
head tests.

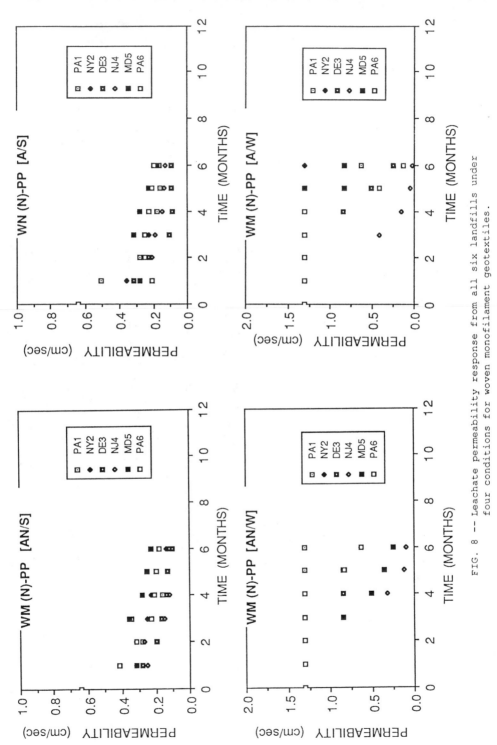

FIG. 8 -- Leachate permeability response from all six landfills under four conditions for woven monofilament geotextiles.

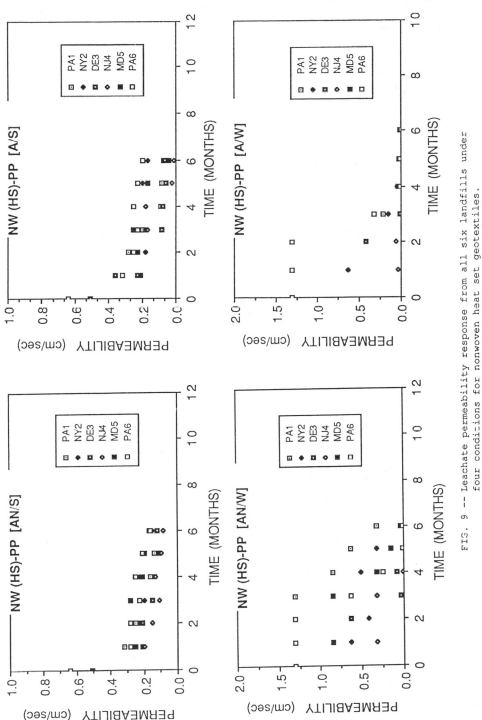

FIG. 9 -- Leachate permeability response from all six landfills under four conditions for nonwoven heat set geotextiles.

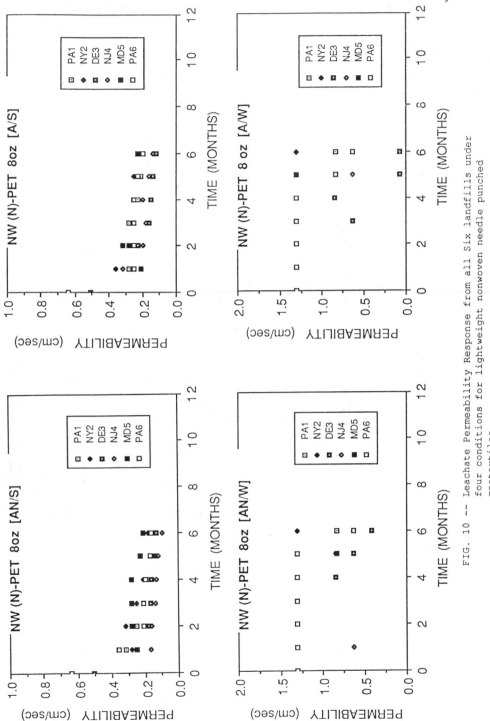

FIG. 10 -- Leachate Permeability Response from all Six landfills under four conditions for lightweight nonwoven needle punched geotextiles.

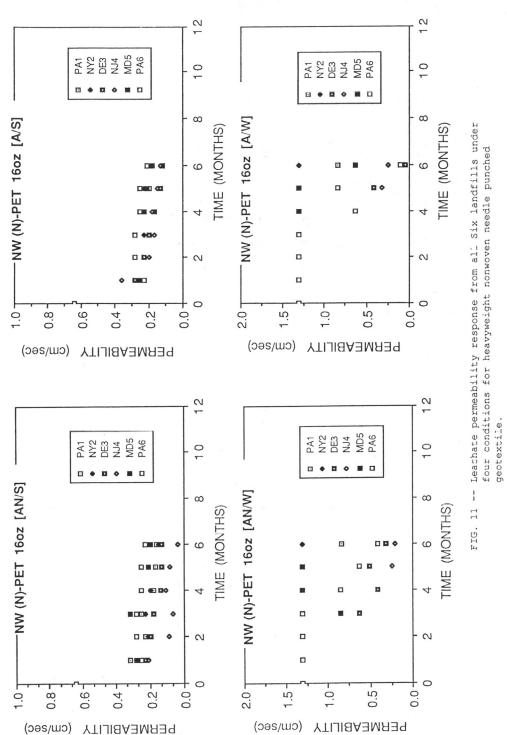

FIG. 11 -- Leachate permeability response from al: Six landfills under four conditions for heavyweight nonwoven needle punched geotextile.

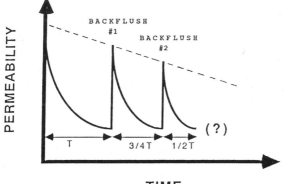

a) Hypothetical Leachate Backflush for Flow Remediation

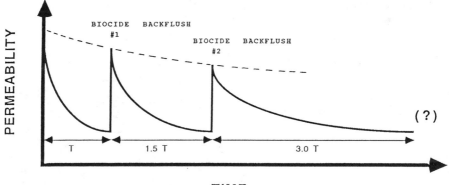

b)   Hypothetical Leachate Backflush with Biocide
      for Flow Remediation

FIG. 12 -- Anticipated flow patterns after repeated backflushing
           trials with leachate (upper curve) and with biocide treated
           leachate (lower curve).

L.G. Tisinger, I.D. Peggs, B.E. Dudzik, J.P. Winfree, and C.E. Carraher, Jr.

MICROSTRUCTURAL ANALYSIS OF A POLYPROPYLENE GEOTEXTILE AFTER LONG-TERM OUTDOOR EXPOSURE

REFERENCE: Tisinger, L.G., Peggs, I.D., Dudzik, B.E., Winfree, J.P., and Carraher, Jr., C.E., "Microstructural Analysis of a Polypropylene Geotextile After Long-Term Outdoor Exposure," Geosynthetic Testing for Waste Containment Applications, ASTM STP 1081, Robert M. Koerner, editor, American Society for Testing and Materials, Philadelphia, 1990.

ABSTRACT: An investigation in two phases was conducted to characterize and evaluate the cause(s) of field deterioration of both black and white nonwoven polypropylene geotextiles in a municipal landfill. A preliminary determination was made in the first phase that the observed decay varied between the materials and was the result of damaging thermal effects and mechanical stress. Observational and analytical data supported but did not definitively establish the causal agents. The second test phase was designed to reproduce the failure mode; it involved thermal exposure of restrained (stressed) and unrestrained samples of black 270 g/m2 (8 oz/yd2) geotextile. An unexposed sample was used to obtain control values. After 72 days of exposure, degradation of the specimens was assessed by measurement of physical and mechanical properties and by scanning electron microscope and microstructural analytical techniques.

KEYWORDS: geotextile, polypropylene, thermal degradation, microanalysis, variability, crystallinity

Mr. Tisinger is program manager, chemistry, and Drs. Peggs and Winfree are president and project manager, respectively, of GeoSyntec, Inc., Boynton Beach, Florida; Mr. Dudzik is a senior project engineer with RMT, Inc., Madison, Wisconsin; and Dr. Carraher is Dean of Science at Florida Atlantic University, Boca Raton, Florida.

INTRODUCTION

Geotextiles and related materials that are used for waste containment applications are selected on the basis of their resistance to chemical, heat, and UV exposure. Such resistance is typically provided through both inherent and additive-supplied (carbon black and antioxidants) properties.

In this case, two nonwoven needle-punched polypropylene geotextiles had been exposed to a northern environment for approximately one year. The geotextiles were located on top of geonet and had not been covered with any type of soil. Figures 1 and 2 show that the geotextiles were severely degraded in discrete areas. One of the two geotextiles contained carbon black, the other (white) material did not. Both geotextiles, however, contained antioxidants. The black geotextile appeared to have degraded more severely than the white.

FIG. 1 -- Site photograph showing localized geotextile thinning.

This two-phase investigation was performed to identify the cause(s) of the degradation and to try to reproduce it. The investigation attempts to determine possible mechanisms of degradation which involves a thermally induced relaxation effect on the fibers.

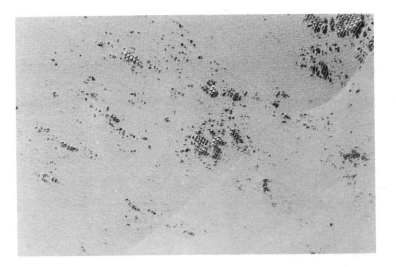

FIG. 2 -- Site photograph of holes in geotextile along the side slope.

It should be noted that leaving geotextiles exposed to the environment is commonly practiced but to varying degrees. In landfills, such exposure most often involves the material that is placed at the top of the slopes.

PROCEDURE AND TEST RESULTS

Phase 1 testing consisted of mechanical and chemical (structural) measurements for comparative evaluation of unused (control) and field (exposed) samples of the two geotextiles. Field material samples were tested which had been removed from areas of serious degradation. Other samples of field materials showed little or no degradation by visual observation.

Phase 1 testing was for fiber strength and elongation, oxidative induction temperature, crystallinity, composition and decomposition temperatures, and scanning electron microscopy (SEM).

While it was learned in Phase 1 that there could be no doubt that degradation in properties as well as appearance had occurred in the field material, the causal agent(s) were not identified.

Phase 2 tests were designed to be more definitive and to simulate the progressive thermal degradation of the black geotextile by attempting to duplicate the observed field behavior of the material using only thermal energy. Samples of unused polypropylene geotextile of one of the same types (black) used in the field were heated in air in a laboratory oven for a given number of days at 100#C. One sample was restrained by the edges in a clamped frame. A second sample was placed unrestrained on a piece of wood which was used to give spatial support and reduce flexing during sample transport. Exposure to UV radiation was omitted from this phase of tests.

Visual signs of gross degradation in the unrestrained sample first appeared at the 33rd day of exposure. Photographs (Figure 3) show the condition of the unrestrained sample after 33 and 72 days of exposure. This degradation progressed through the exposure period but appeared to affect only one side of the sample. The restrained sample was visually unaffected by the thermal exposure.

FIG. 3 -- Unrestrained geotextile sample after 33 days (left) and 72 days (right) of thermal exposure at 100°C.

The weight and dimensions of the unrestrained sample were measured at regular intervals. Mass and dimensional data are plotted against time of exposure in Figure 4. There appears to be an induction period of 20-25 days before appreciable weight losses are observed. The weight loss appeared to result from mechanical breakage and loss of fibers from the geotextile sample during specimen handling. The weight loss becomes less accelerated around day 50, possibly the result of the failure mechanism reducing the geotextile to a nominal and minimum value.

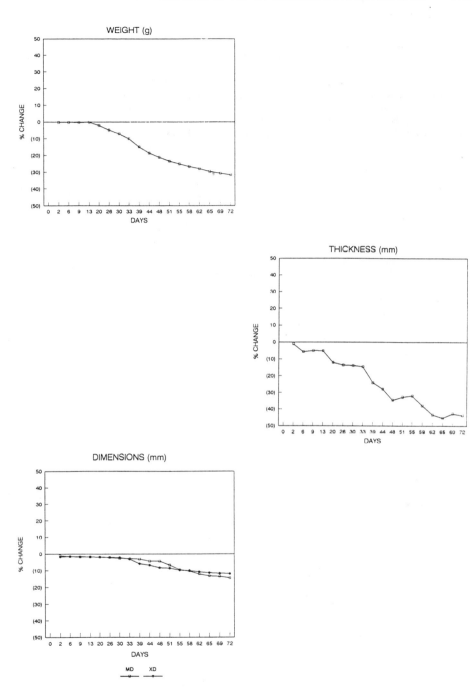

FIG. 4 -- Weight and dimensional changes of unrestrained black polypropylene geotextile samples.

In Phase 2 as in Phase 1, individual fibers were carefully removed from the control sample, in this case from as close to the damaged area of the unrestrained sample as possible and from a randomly selected area of the restrained sample. Tensile tests were performed on each fiber according to ASTM standard test method D3822. The force and elongation at failure were monitored.

Differential scanning calorimetry (DSC), thermal gravimetric analysis (TGA), and infrared (IR) spectrophotometry were used to examine the control, restrained, and unrestrained geotextile samples. Specimens cut from the least and most degraded areas of the unrestrained sample were examined. DSC and TGA results were obtained using a Perkin-Elmer DSC-4 and TGS-2, respectively, both equipped with a system IV and model 3700 data acquisition and processing facility. IR spectra were obtained using a Perkin-Elmer 283B infrared spectrophotometer with a model 3500 data station.

The results of the Phase 2 tests are shown in Tables 1 through 3. Tensile strength and elongation of single fibers are compared in Table 4.

TABLE 1 -- Degree of Crystallinity of Black Geotextile Showing
the Effects of 72-Day Thermal Exposure
Determined by DSC Measurements

Sample	Unexposed (%)	Exposed (%)	Change (%)
Control	32.99	--	--
Restrained	--	33.31	1.00
Unrestrained			
Least Degraded Area	--	34.36	4.10
Most Degraded Area	--	44.00	33.00

The degree of crystallinity measurements show some change between unexposed control and physically least degraded material, both restrained and unrestrained. The most degraded portion of the unrestrained material exhibited a 33 percent increase in degree of crystallinity.

TABLE 2 -- Oxidative Induction Temperatures of Black Geotextile
Showing the Effects of 72-Day Thermal Exposure
Determined by DSC Measurements

Sample	Unexposed (°C)	Exposed (°C)	Change (%)
Control	210.5	--	--
Restrained	--	213.2	1.3
Unrestrained			
Least Degraded Area	--	210.3	-0.1
Most Degraded Area	--	214.9	2.1

The oxidative induction temperature (OIT) values show both that
additives were evenly distributed throughout the geotextile (since
differences in OIT among the samples were insignificant) and that
exposure to heat did not significantly affect oxidative stability of
the geotextiles (exposed samples displayed little change from the
control value). Negligible differences were found in infrared spectra
between the control and thermally exposed samples. Infrared bands
that may be associated with oxidation were observed in the infrared
spectrum of the unrestrained geotextile.

The greater changes in crystallinity (increase), mechanical
properties (decreases), and OIT (increase) for unrestrained samples is
reasonable, consistent with an increased crystallinity for the
unrestrained material (1,2). Density increases with increased
crystallinity (for instance, the density of amorphous PP is about
0.85-0.89 whereas the density for crystalline isotactic PP is about
0.92 to 0.94 (2)). The stress present for the restrained material
encourages torsional restriction, chain slippage, etc. but discourages
attempts by the chains to pack more closely. The unrestrained sample
is free to take advantage of the added thermal energy by becoming more
crystalline.

The TGA test results are consistent with those of DSC tests for
crystallinity. The crystallinity value increased significantly for
the unrestrained sample. Such an increase is consistent with
secondary crystallization of the polymer. Secondary crystallization
increases thermal stability, thus decomposition temperatures increase
as a result. Conversely, the polymer chains comprising restrained
geotextile fibers have reduced mobility, therefore increases in both
crystallinity and decomposition temperatures are less significant.

TABLE 3 -- Composition and Decomposition Temperatures of Black
Geotextile Showing the Effects of 72-Day Thermal
Exposure Determined by TGA Measurements

Property	Unit	Sample			
		Control	Restrained	Unrestrained Least Degraded	Unrestrained Most Degraded
Volatiles	%	0.23	0.20	0.13	0.03
Polymer	%	97.31	96.41	97.28	96.54
Residue	%	2.46	3.39	2.59	3.43
Temperature at 5% Weight Loss	°C	251.34	271.95	264.46	304.73
Temperature at 50% Weight Loss	°C	338.44	363.73	363.73	404.93

TABLE 4 -- Tensile Test Results for Single Fibers of Black
Geotextile Showing the Effects of 72-Day
Thermal Exposure

Sample	Unit	Mean	Percentage Change
Control			
Ultimate Load	kN	0.22	--
Elongation	%	18.00	--
Unrestrained			
Ultimate Load	kN	0.15	-31.2%
Elongation	%	10.40	-42.2%
Restrained			
Ultimate Load	kN	0.17	-21.3%
Elongation	%	12.80	-28.9%

Scanning electron microscopy (SEM) was performed to define the
macrostructural characteristics of the degradation process from the
features of the fiber fracture faces and outside surfaces of the
fibers. Specimens removed from the unexposed control sample, from the
most damaged areas of the unrestrained sample, and from a randomly
selected area of the restrained sample were examined. In preparation
for examination of the specimens by SEM, all specimens were coated at
the same time with a gold-palladium alloy to provide the electrical
conductivity required for SEM examination.

The SEM analyses performed in Phases 1 and 2 showed that unexposed control fibers of the black geotextile displayed an almost featureless, smooth surface as shown in Figure 5. In comparison, the fibers removed from the field showed both circumferential cracking (Figure 6) and circumferential wrinkling combined with longitudinal splitting (Figure 7). The wrinkling has the appearance of a loose "sheath" being gathered up on a rigid core. However, the cross sectional fracture faces of the fibers do not indicate that there is a sheath/core structure. The circumferential wrinkles appear, from Figure 7, to be related to circumferential voids just underneath the surface of the fiber. In the Phase 1 study, the defects on the field sample fibers were not uniformly distributed throughout the specimens but occurred in more or less discrete locations, and then only on individual fibers.

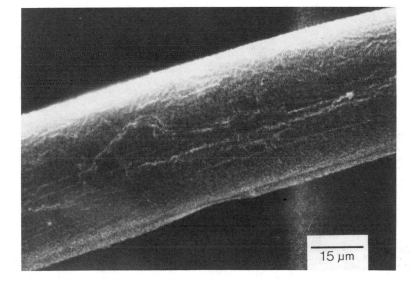

FIG. 5 -- Smooth surface of unexposed fiber (Phase 1).

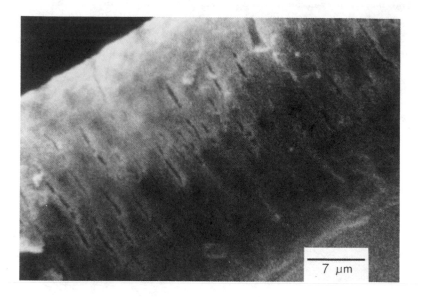

FIG. 6 -- Circumferential cracking in exposed fiber (Phase 1).

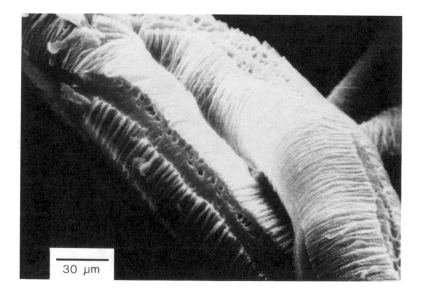

FIG. 7 -- Longitudinal fissures and wrinkling in surface of exposed
            fibers (Phase 1).

The cracking observed in Figure 6 is the early stage of the severe cracking typically observed in polypropylene fibers that have been intentionally exposed in long-term weathering tests to UV and thermal radiation, as shown in Figure 8.

25 μm

FIG. 8 -- Extreme circumferential cracking in UV unprotected polypropylene fibers exposed for three months (for illustration purposes only).

Fibers removed from the unrestrained specimen in the Phase 2 SEM examination showed relatively clean, homogeneous fracture characteristics (Figure 9) with some evidence of circumferential cracking on the surface. Again, these features were not widespread. In the more deteriorated areas of the unrestrained sample, the fibers showed little wrinkling (Figure 10) but a considerable amount of fiber damage. Fibers had fractured (Figure 11) often with a ligament extension to the fracture face at one area of the circumference.

The surface of the fibers also appeared to be in the early stages of wrinkling, and in several locations, there was evidence of distortion (kinks) in the fibers (Figure 11) as though an inside "core" of the fiber had broken and an outside sheath had crumpled. It appears that the fiber then breaks but that ligaments at or near the surface (possibly the sheath) may still hold the ends together (Figure 12). This is demonstrated clearly in Figure 13 which shows several breaks in a single fiber that are held together by the ligaments. A specimen removed from a seemingly less damaged area of the restrained sample also showed two broken but not separated fibers (Figure 14).

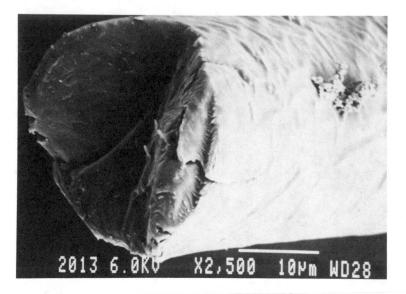

FIG. 9 -- Brittle fracture and cracks on side of fiber (unrestrained sample, Phase 2).

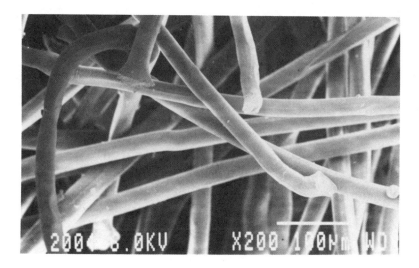

FIG. 10 -- Smooth fibers showing expanded, fractured ends (unrestrained sample, Phase 2).

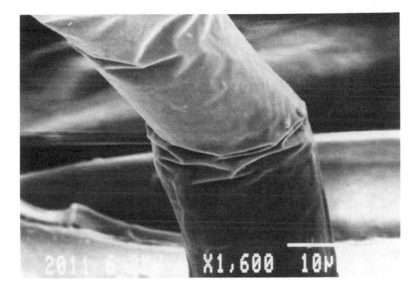

FIG. 11 -- Internally fractured, externally crumpled fiber (unrestrained sample, Phase 2).

FIG. 12 -- Fractured fiber held together by outer layers (unrestrained sample, Phase 2).

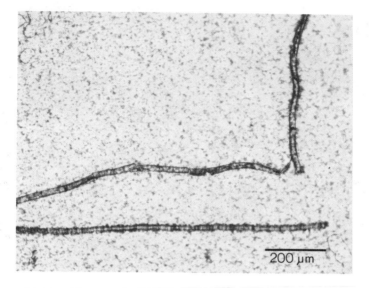

FIG. 13 -- Fiber broken into many segments held together by surface
ligaments (unrestrained sample, Phase 2).

FIG. 14 -- Two fractured but unseparated fibers (restrained sample,
Phase 2).

In Phase 2 many of the unexposed control fibers and restrained specimen fibers showed fiber ends that were expanded, wrinkled, and longitudinally split as noted in Figure 7 for the Phase 1 field samples. These features appeared to be somewhat anomalous, and light microscope examination of the samples prior to preparation of the SEM specimens clearly showed that the large number of wrinkled, bulbous fiber ends was not present. It was therefore concluded that the coating process had introduced these features to the fibers. This could be done by overheating or by exposure to UV radiation possibly generated within the chamber as the coating was deposited on the fibers.

While not exhaustive, the microscopical study indicates that there are three distinct physical stages in the thermal degradation process. In the first stage, the inside of the fiber actually fractures and the outer surface begins to crumple. In some cases, the outer surface begins to crack and the whole fiber breaks. In the second stage, with the additional absorption of thermal energy (and perhaps UV radiation), the crumpling of the outer surface at broken ends develops into bulbous wrinkling with longitudinal splits. In the final stage, the severely deformed ends break off. Only when the ends break off does the geotextile begin to lose weight and visibly degrade.

More analytical and microscopical work is required to elucidate this complex phenomenon.

DISCUSSION

This investigation, while not exhaustive, sheds substantial light on how the degradation of polypropylene geotextiles occurred in service by providing insight to the mechanism(s) of decay of such products.

Laboratory simulation of field conditions without light (UV) effects allowed for an examination of a process of thermal degradation of properties to a near total loss of integrity of test material.

The thermally induced degradation was obviously not accelerated by restricting the geotextile movement, contrary to what was suspected from the Phase 1 observations.

Even more surprising, perhaps, was finding that "pockets" or localized areas of more thermally sensitive material of varying size and distribution occur in random fashion throughout the geotextile. Such pockets are likely due to variability in the extent of fiber orientation distributed throughout the sample.

The cause of the circumferential cracking and ultimate fiber breaking is either chain scission breaking the long polymer molecules oriented parallel to the axis of the fiber or embrittlement due to secondary crystallization. Small surface breaks provide access of oxygen to the inner portions of the fiber and therefore promote circumferential and radial crack propagation.

Generally, the microscopical observations and theories match the weight loss observations. As the fibers initially fracture, but are held together by surface ligaments, there is no weight loss, but as the wrinkled ends break away, a loss in weight begins to occur.   The difference in performance between the restrained and unrestrained samples indicates that the induced orientation in the restrained sample reduced the susceptibility of the geotextile to secondary crystallization.   Conversely, the unrestrained geotextile suffered significant degradation, both visually and microscopically.   This difference appears to indicate that the primary causal factor of the degradation observed both in the laboratory and in the field is likely attributable to thermally induced relaxation of the fibers, leading to secondary crystallization induced embrittlement of the fibers.

Both the unrestrained and the field samples suffered damage in localized pockets.   Initial work suggested that antioxidants were not evenly distributed throughout the sample.   However, the oxidative induction temperature tests showed equivalent values throughout the samples.   Therefore, it appears that antioxidant distribution was uniform in the samples.   A more plausible explanation may be that orientation of the fibers is not equally distributed throughout the geotextiles, thus areas that are more highly oriented (fibers pulled more tightly) may be less susceptible to crystallization.

In summary, two major changes are known to occur as polypropylene is exposed to radiation (light and thermal (3,4)).   These major changes are increased crosslinking and increased crystallization. Both factors give a decrease in the mechanical properties of polypropylene, eventually leading to stress fracture.   Both factors are present in landfills (Phase 1) and must be contended with.   (Light radiation factors are minimized for Phase 2 studies.)   In the present study, the presence of suitable additives appears to minimize the formation of crosslinks as a major cause of polypropylene geotextile failure.   Unfortunately, while the melting temperature is well above 100°C, the short-term, whole chain mobility is restricted under a normal environment.   Thermally induced local chain movement is sufficient to permit long-term, whole chain movement which leads to crystallization.   Such crystallization may be retarded through a careful, controlled introduction of crosslinking such that whole chain mobility is discouraged, but where mechanical properties are not adversely affected.

CONCLUSIONS

Based on this study, it is tentatively concluded that the fibers in the polypropylene geotextiles have degraded due to thermally induced relaxation and secondary crystallization of fibers which resulted in significant reduction in tensile strength and ductility and loss of integrity.

The degradation across the geotextile appears to be nonuniform, with areas of failure located adjacent to fully functional material. This is most probably due to inhomogeneity in the manufacturing process, producing regions of variable fiber orientation.

REFERENCES

1. Seymour, R., and Carraher, C., Polymer Chemistry, Dekker, New York, 1989.

2. Lieberman, R., and Barbe, P.C., _Encyclopedia of Polymer Science and Engineering_ (eds. H. Mark, N. Bikales, C. Overberger, G. Menges and J. Kroschwitz) 2nd ed., John Wiley and Sons, New York, Vol. 13, 1988.

3. Carlsson, D.J., and Miles, D.M., _J. Macromol. Chem._, C14 (1), 65 (1976).

4. Severini, F., _Chim. Ind. Milan_, 60(9), 743 (1978).

Performance Behavior of Several Geosynthetic Systems

Nandakumaran Paruvakat, Gerald W. Sevick, John Boschuk Jr., and Steven Kollodge

DESIGN AND TESTING OF A LANDFILL FINAL COVER WITH GEOMEMBRANE

REFERENCE: Paruvakat, N., Sevick, G.W., Boschuk, J., Jr., and Kollodge, S., "Design and Testing of a Landfill Final Cover With Geomembrane", Geosynthetic Testing for Waste Containment Application, ASTM STP 1081, Robert M. Koerner, editor, American Society for Testing and Materials, Philadelphia, 1990.

ABSTRACT: In order to evaluate the feasibility of construction and the adequacy of design of a landfill final cover incorporating a geomembrane, a three-phase investigation was conducted. The frictional characteristics of a sand-geomembrane interface was determined using a tilt apparatus. Stability of an element of the final cover under the design rainfall and freeze-thaw conditions was evaluated using a large size tilt table. Two test plots constructed at 25 percent slopes and periodic observations for over a year provided information on construction feasibility and the behavior of the final cover system under field conditions. Based on the results of the three phases of investigations it was concluded that final covers incorporating geomembranes can be designed and constructed for slopes of 25 percent.

KEYWORDS: landfill, final covers, geomembranes, stability, friction tests, laboratory tests, field tests

Dr. Paruvakat is Section Manager and Mr. Sevick is Group Vice President at Foth & Van Dyke and Associates Inc., 2737 S. Ridge Road, Green Bay, WI 54307; Mr. Boschuk is President of J&L Testing Co., Inc., 938 S. Central Avenue, Canonsburg, PA 15317; Mr. Kollodge is Senior Site Engineer at Anoka Regional Sanitary Landfill, 14730 Sunfish Lake Boulevard, Anoka, MN 55303.

INTRODUCTION

The investigations reported in this paper pertain to Anoka Regional Sanitary Landfill at Anoka, Minnesota and was performed during 1988-90. In order to reduce leachate generation, the final cover designed for the landfill incorporated a composite hydraulic barrier with a geomembrane overlying a compacted lime sludge (the compacted lime sludge was later replaced with a compacted clay layer). A drainage layer over the geomembrane and a rooting layer and topsoil over the drainage layer are also included in the final cover.

A cross section of the cover system showing the different components is given in Figure 1.

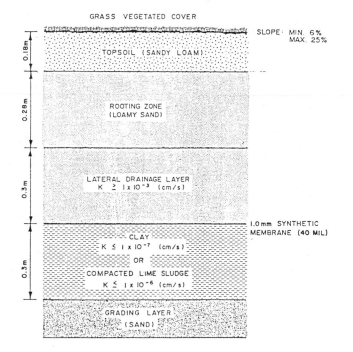

FIG. 1 -- Cross section of final cover.

Final Cover System Design

The slopes of the final cover ranged between a minimum of 6 percent and a maximum of 25 percent with a majority of the slopes at 25 percent.

The geomembrane in a composite final cover system needs to have adequate resistance to chemicals, vapors, ozone, ultraviolet rays, and mechanical, plant and animal damage. Adequate coefficient of friction between the membrane and the drainage layer is necessary to achieve satisfactory cover soils stability. Even though high density polyethylene (HDPE) has many advantages, the non-textured sheets are smooth and pose stability concerns for a steep final cover system. Textured HDPE which has better frictional properties is a comparatively new product and its ability to accommodate potentially large landfill settlements has not yet been demonstrated. Another relatively new product which shows promise for this application is very low density polyethylene (VLDPE). Two especially attractive features of this product for landfill cover application are its flexibility and good frictional characteristics. Both materials, namely a textured HDPE (Gundline HDT manufactured by Gundle) and a VLDPE (Duraflex manufactured by Poly-America), were included in the feasibility investigations.

The drainage layer consists of clean sand with a hydraulic conductivity not less than 10^{-3} cm/sec. The vegetated top layer consists of a rooting layer and a topsoil layer selected on the basis of adequate moisture retention capacity.

Diversion berms and drainage ditches at approximate vertical spacing of 12 m (40 ft) were provided to reduce the erosion potential of the cover soils. The flow in the drainage ditches is transported downslope into the perimeter ditches via HDPE pipes. The diversion berms and drainage ditches also serve another important function. The water flowing through the drainage layer discharges into the ditch where the sand layer is daylighted through a stone and geotextile filter pack. This detail is shown on Figure 2.

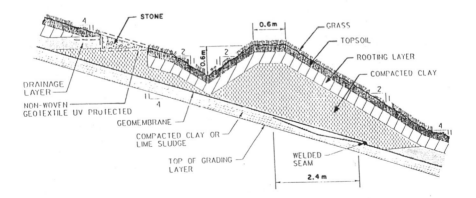

FIG. 2 -- Drainage ditch detail.

By thus letting the water in the drainage layer to discharge at several intervals, the pore pressures on the potential slip plane are relieved and the stability of the cover soils is greatly enhanced.

Need for the Investigations

The investigations described in this paper were performed primarily to demonstrate that the cover system is constructible and stable under actual site conditions. Though the analysis of the proposed cover system's stability using available information and engineering judgement indicated that a sufficient factor of safety is provided by the design, the relatively new technology, new materials proposed for use, and the non-availability of a significant performance database warranted a detailed investigation. Evaluation of the relevant material properties for use in the stability analyses and construction of test plots were considered necessary to provide the answers to several questions. A detailed laboratory evaluation of the stability of the cover soils resting over the geomembrane was dictated by the practical limitation on the duration of the study of the test plots. During this time period the worst loading conditions with regard to rainfall and freeze-thaw may not occur. The laboratory tests were also considered necessary to evaluate the reserve resistance available in the system as designed, in terms of the factor of safety under rainfall conditions.

PHASE I INVESTIGATIONS - FRICTION TESTS

As mentioned earlier, an initial screening process identified two potential geomembranes for use as the primary hydraulic barrier of the final cover. Phase I studies were performed to evaluate the frictional characteristics of the membrane-sand interface. The relatively low stress levels associated with the landfill final covers can be handled much more easily in a tilting apparatus. Thus, this apparatus was preferred over the conventional direct shear apparatus. Further, at very low confining pressures, the failure surface is typically between soil and the geosynthetic membrane since little interlocking occurs [1]. Because of the above considerations, a tilt apparatus where the failure surface is limited to the soil-synthetic interface, was used. Figure 3 shows a schematic of the apparatus.

Sample preparation consisted of the construction of one component of the interface (the geomembrane in this case) on a hydraulically activated tilt table and the construction of the other component (sand or other construction material) on the base of a sliding box 0.45 m x 0.45 m (18 in x 18 in) resting on the table. The sliding box is capable of being loaded with custom-built steel plates to create low stress intensities (less than 48 kPa) at the interface. The test procedure used was as follows. The tilt table was kept horizontal and the desired contact stress of 4.8 kPa (100 psf) was applied. The table was then inclined at a rate of approximately 3.5 centiradian (2°) per minute until a slight initial movement of the box was noted. The angle of inclination was then further increased until the box fully slid. The friction angle used in evaluation of Phase II tests corresponded to the one at full sliding. The initial slip is believed to be due to minor adjustments in the soil. No movement of the box is observed in between the initial slip and full sliding. The test was repeated at least three times to check the reproducibility of results. The test procedure was repeated at two more contact stresses of 9.6 and 14.4 kPa (200 and 300 psf). Two test modes, ie., under dry conditions and with water flowing vertically through the sand, were also used. The test results are shown in Table 1. On the basis of these results, VLDPE was chosen as the primary hydraulic barrier for Phase II of the investigation, ie., tilt table test of a full scale element of the final cover. The main considerations influencing the selection of VLDPE was its better flexibility and practically similar frictional properties as the textured HDPE. The tests also showed that the minimum friction angle (and hence potentially the

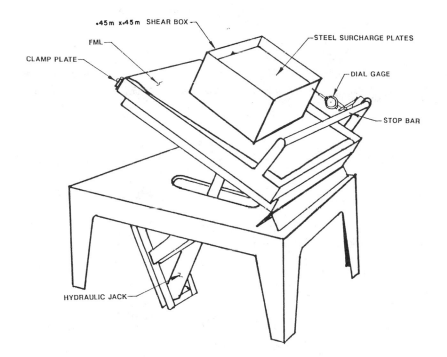

FIG. 3 -- Tilt apparatus for friction tests.

failure plane in the cover system) was between the geomembrane and the drainage sand above it. The tests additionally provided data for analyses of the Phase II test results where the cover element was housed in a bin with acrylic sidewalls.

PHASE II INVESTIGATION, TILT TABLE TESTS OF A FULL SCALE FINAL COVER ELEMENT

    The test apparatus consisted of a tilt table and a bin in which a full scale element of the final cover was constructed. The apparatus is shown in Figure 4.

    The bin measures 3.66 m (12 ft) x 1.52 m (5 ft) x 0.91 m (3 ft) high and consists of welded steel frame construction. The base was constructed of a 6.35 mm (0.25) channel reinforced steel plate with a 0.6 m (2 ft) drop section at the lower end to accommodate a drainage collection system. The bin was seated on a modified dump truck bed which was cased in a 3.66 m x 3.66 m x 0.3 m (12' x 12' x 1') reinforced concrete slab for stability. The two sidewalls of the bin consisted of a steel frame and 1.27 cm (0.5 in) thick plexiglass panels.

TABLE 1--Friction tests using tilt apparatus

Test No.	Interface Materials		Normal Stress kPa	Inclination in Radian (Deg) at	
	Lower Component	Upper Components[a]		Initial Slip	Full Sliding
1	VLDPE	Satur. Sand	4.8	0.384(22)	0.454(26)
2	VLDPE	Satur. Sand	9.6	0.384(22)	0.454(26)
3	VLDPE	Satur. Sand	14.4	0.384(22)	0.454(26)
4	VLDPE	Dry Sand	4.8	0.436(25)	0.48(27.5)
5	VLDPE	Dry Sand	9.6	0.436(25)	0.471(27)
6	VLDPE	Dry Sand	14.4	0.436(25)	0.471(27)
7	HDT	Satur. Sand	4.8	0.436(25)	0.48(27.5)
8	HDT	Satur. Sand	9.6	0.419(24)	0.48(27.5)
9	HDT	Satur. Sand	14.4	0.436(25)	0.48(27.5)
10	HDT	Dry Sand	4.8	0.436(25)	0.52(29.8)
11	HDT	Dry Sand	9.6	0.436(25)	0.53(30.4)
12	HDT	Dry Sand	14.4	0.436(25)	0.52(29.8)
13	HDT	Lime Sludge	4.8	0.436(25)	0.768(44)
14	HDT	Lime Sludge	9.6	0.445(25.5)	0.768(44)
15	HDT	Lime Sludge	14.4	0.454(26)	0.768(44)
16	Acrylic	Dry Sand	4.8	No Warning	0.41(23.5)
17	Acrylic	Dry Sand	9.6	No Warning	0.419(24)
18	Acrylic	Dry Sand	14.4	No Warning	0.419(24)
19	Acrylic	Lime Sludge	4.8	0.436(25)	0.576(33)
20	Acrylic	Lime Sludge	9.6	0.436(25)	0.585(33.5)
21	Acrylic	Lime Sludge	14.4	0.436(25)	0.585(33.5)
22	Acrylic	Rooting Soil	4.8	0.436(25)	0.471(27)
23	Acrylic	Rooting Soil	9.6	0.436(25)	0.471(27)
24	Acrylic	Rooting Soil	14.4	0.436(25)	0.471(27)

[a]Notes: Dry density of sand in tests - 2095.5 kg/m^3 (130.8 pcf).
Dry density of lime sludge 923.8 kg/m^3 (57.6 pcf).
Dry density of rooting soil 1915.9 kg/m^3 (119.6 pcf).

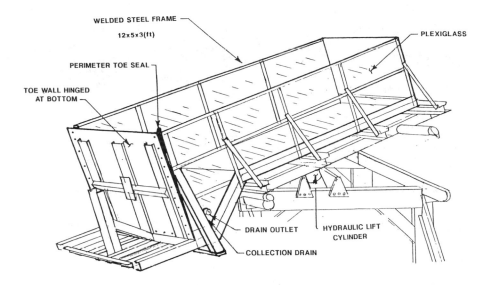

FIG. 4 -- Tilt table apparatus.

The inner sides of the sidewalls were covered with 3.2 mm (0.125 in) thick acrylic to minimize friction between the cover materials and the sidewalls. The back (up-slope) wall of the bin consisted of a 6.4 mm (0.25 in) thick channel-reinforced plate with a slit at the bottom to allow protrusion of the FML for monitoring tension. At this end a rigid harness assembly was constructed to house a 88.96 kN (20,000 lb force) load ring and steel clamp plate to hold the FML for monitoring tension. The toe (downslope) wall was constructed of 6.4 mm thick channel-reinforced steel plate hinged at the bottom to allow rotation of the wall. The rotating toe wall was incorporated in the apparatus in an effort to minimize the effects of a rigid down slope wall. By providing a hinged wall and supporting it by a retractable screw, the end wall reaction could be brought to the minimum possible value. The active thrust which is the minimum reaction from the end wall was identified by monitoring the force on the wall when it was retracted. The perimeter of the toe wall was made watertight using a strip of polyethylene connected to the wall as well as the sidewalls and base of the bin. The collection drain at the toe of the model was provided with a ball value so that blocked-drainage conditions could be simulated.

The table was fitted with a pipe hinge at the toe end of the support frame. The table could be tilted using a 150 mm (6 in) diameter telescoping hydraulic cylinder activated with a 250 watt, 120 volt fixed speed motor. In addition, a channel support with pre-drilled holes at selected locations and a pin was used to hold the table at each of the prescribed inclinations for the tests.

Instrumentation

The instrumentation employed for the test are described in this section. Geosynthetic movement monitoring pins consisting of a sheet metal pointer stapled to the geomembrane and sliding against the sidewall inside a plexiglass housing were used to monitor potential movement/elongation of the membrane. A steel scale was placed outside of the sidewall as a reference point. Eight such pins were employed.

Vertical sand column indicators were incorporated in the cover element during construction to monitor shear and creep movements in the element during the tilting test. These were placed along the side of the bin and extended perpendicular to the membrane all the way up to the surface of the element. These columns were constructed at 11 locations on one side and 12 locations on the other.

Water level indicators were used to monitor the water level within the cover element during different times of tilt test.

Six thermocouples were employed to monitor the temperature of the cover soils during the freeze-thaw test.

Materials and Construction of the Cover Element

Coarse gravel (75 mm maximum size) was used for the construction of the collection drain. This gravel was hand-compacted to provide a firm base for the overlying compacted lime sludge. A heat bonded non-woven geotextile was used as a separator between the collection drain and the soil/sludge above. After filling the drain recess with gravel and placing a geotextile over it, the base of the bin was covered with lime sludge and compacted, but for an area of 0.3 m x 1.52 m near the toe. The compacted thickness of the lime sludge layer was 75 mm. The geotextile left uncovered by the lime sludge in the 0.3 m x 1.52 m area was then folded back over the lime sludge. The geomembrane which had been precut to be slightly wider than the width of the bin was then spread over the lime sludge and trimmed so that the edges conformed exactly to the base at the sidewalls. The toe end of the liner was folded down over the face of the underlying lime sludge. The up-slope end of the membrane was extended through the slot in the back wall and secured to the tension harness after the soil placement was completed.

Before placing the drainage layer and the rooting soil, compaction tests were performed. On the basis of these tests, the required compactive efforts to achieve the design densities were calculated. Manual compaction in 0.1 m (4 in) thick lifts were employed.

A piece of non-woven geotextile was placed over the exposed drainage stone along the toe wall. The required drainage layer was constructed over this area and the geomembrane. The 0.3 m x 1.52 m area near the toe wall thus became an inlet for water inflow into the collection drain.

The sand layer and the rooting layer were compacted and placed as per the design shown in Figure 1. The topsoil was not compacted. A burlap cover was placed over the topsoil to simulate vegetative cover.

Freeze-Thaw Test Procedure

Test 1: thaw from bottom up: The first freeze-thaw test entailed freezing the entire slope element, after being presaturated, inclining the table to 25 percent slope and then thawing the table from the bottom up. The purpose of this test was to evaluate if pore water pressure build up occurs at the soil-FML contact when thawing takes place at this contact plane.

Prior to freezing, the table was saturated in order to simulate "worst-case" conditions. The toe drain was open during saturation. The proving ring reaction on the toe wall was brought to the initial "at-rest" value of 20.24 kN, (4,550 lb) (determined during soil placement) while the table was in the horizontal position. In addition, the geosynthetic liner was pre-tensioned to 1.98 kN (445 lb) prior to freezing the table in order to pull the slack out of the exposed portion of the liner.

Dry ice was used to reduce the soil temperatures to below freezing. Six thermocouples were monitored to document the temperature inside the model. The blocks of dry ice were placed on top of the table and covered with 50 mm thick polystyrene insulation and reinforced plastic. Subsequently, thin blocks of dry ice, supported on plywood sheets, were placed around and underneath the table to accelerate the freezing process. The tilt table was inclined to 25 percent slope after freezing was completed.

In order to thaw the soils from the bottom up, a canopy of reinforced polyethylene was draped around the table frame to the ground. A kerosene torpedo heater was positioned under the canopy, directing a blanket of heat along the bottom of the table. Adjustments were made in the location and direction of the heater based on frequent temperature readings to insure an even thawing process along a plane parallel to the base of the table.

In the absence of a suitable pore pressure transducer to monitor instantaneous increases at the liner interface, close monitoring of the toe wall reaction during thaw was performed to obtain qualitative information on such an occurrence. An increase in the toe reaction should result when transfer of load from the liner/sand interface occurs due to reduction in effective stresses. In the test described here, no increase in toe reaction was noted. No soil movements were observed even after the thawing had progressed sufficiently through the sand layer. This observation is considered indicative of sufficiently quick dissipation of pore pressures due to drainage.

Test 2: thaw from top down: Following the previously discussed freeze thaw test, the table was lowered to the horizontal position. The cover soils were refrozen, inclined to 25 percent slope and allowed to thaw from the top down due to natural ambient conditions. The purpose of this test scenario was to evaluate whether or not this process would cause slope movements due to the softening of soils during thaw.

The "at-rest" toe reaction of 20.24 kN was re-applied to the table prior to lowering the unit to the horizontal position from the 25 percent inclination of Test 1. Dry ice was placed beneath the table to re-freeze the system. Ambient temperatures, which were generally slightly below freezing, insured the remainder of the model remained frozen. At a 25 percent slope inclination, the cover element was allowed to thaw naturally as ambient temperatures rose. A canopy of black fabric was draped around the bottom of the tilt table to prevent thawing to occur from the bottom up. (The elevated table bottom was exposed to direct sunlight in the late afternoon and evening). Temperature probes were used to monitor the progress of thaw.

The toe reaction was continuously monitored during the thawing process in order to detect any load transfer which might occur. The reaction increased to 21.20 kN (4,765 lb) immediately after inclining the table to 25 percent. Thereafter, no changes were observed even after total thawing.

Rainfall Simulation Test Procedure

This phase of the tilt table testing entailed fitting the unit with a circulating water system capable of simulating a 24-hour, 25-year storm event for this geographic area. The rainfall simulation tests were performed at 25, 30 and 35 percent slopes. A hydrograph was developed from the estimated 25-Year Storm of 115 mm in 24 hours.

The storm water system consisted of a supply/recirculation basin, pump, flow meter, low-flow cut-off, sprinkler arrangement, collection gutter, and runoff collection drum. The system for handling the runoff was designed on the basis of estimates using HELP model. During the tests heavy buildup of water at the toe and clogging of the plumbing was observed. The system was subsequently redesigned to handle the increased flow.

The tests consisted of the following steps:

- Saturate the model to the top of the topsoil with the table horizontal.
- Open drain pipe until all water is drained from table.
- Slowly incline the table to the specified test slope.
- Simulate 24-hour rainfall.
- After completion of rainfall test, close drain pipe, continue the rainfall simulation and allow water to rise to level of topsoil at the toe. Two burettes installed along the side of the table were used to monitor the water levels within the cover soils. This step was included to simulate the case of blocked drainage.

In the 25 percent slope, test, the last step was not included.

After completion of the rainfall simulation test at 35 percent slope, the table was slowly tilted to increase the slope, with the rainfall system running at the peak flow of 30×10^{-5} cubic meter per second (7 gpm). Toe wall and liner reactions, as well as movement of geosynthetic pins, were monitored at different inclinations of the table.

The toe reaction was continuously adjusted to the lowest achievable value at frequent intervals during the test in an attempt to minimize the effects of the unrealistic resistance. The "failure" slope was determined from the observations of toe reaction and movements of the geosynthetic pins and the vertical sand columns to be between 56 and 59 percent. The process of raising the table was, however, continued to a slope of 68 percent in order to fully develop the failure plane visible along the sides of the model. Figure 5 shows the failure plane near the up slope end where it was inclined at about 29 degrees with the horizontal. A major portion of the failure plane was along the upper surface of the membrane.

FIG. 5 -- Failure surface in the cover soils.

Analysis of Test Results

The observations during the tilt test showed that the cover soils slipped practically along the membrane as a block. Thus, the large scale element test confirmed the assumption in the earlier analyses that the potential failure surface within the final cover system was the plane along the geomembrane-drainage sand interface. Further assumptions used in the analyses included those regarding the pore pressures and the mobilized coefficient of friction between the sand and the membrane. It was, therefore, considered prudent to analyze the behavior of the large scale element in the tilting table experiment using a sliding block analysis (infinite slope model) and a finite element analysis. The calibration of the stability analysis models as described above was performed by computing the factors of safety of the element against sliding at various inclinations. For the finite element method, the toe resistance was determined from the measured values. With both models, two cases each were analyzed, one without any side forces and one using the ultimate side resistance in the tilt table test. For compiling this side resistance, the observed friction angle between the model materials and acrylic was utilized. In the analyses, the mobilized friction angle between the geomembrane and the sand was taken as the ultimate friction angle measured during Phase I of the tests. The results obtained are shown in Figure 6.

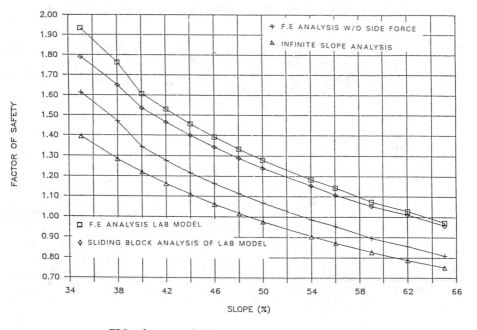

FIG. 6 -- Stability analysis results.

The two curves identified as analysis of lab model in Figure 6 were developed using all the forces present in the model except the pore pressures. The forces considered included the measured toe resistance and the sidewall friction. With these conditions both the analytical models, namely the infinite slope analysis and the finite element method predicted failure of the lab model at a slope of 63 percent, slightly higher than the observed failure slope of 56 to 59 percent. This difference is believed to be due to possible reduction of sliding resistance caused by pore pressure.

The pore pressures were not included in the analysis on the assumption that the drain outlet at the toe will cause the pore pressures to reduce to zero. The observations suggest that the pore pressure relief has been of the order of 80 percent as surmised from the observation of the failure slope which is about ten percent less than the calculated value. Because these appear extremely reasonable, the conclusions drawn from this analysis are that both the analytical methods used are adequate to predict the stability of the cover soils and that the toe drain causes most of the pore pressures to reduce to zero near the toe.

PHASE III - FIELD DEMONSTRATION TEST PLOTS

As discussed earlier in this paper, two test plots each measuring approximately 58 m (190 ft) long and approximately 14 m (45 ft) wide were constructed at the Anoka Regional Sanitary Landfill. The two plots were identical except for the difference that a Gundline HDT membrane was used in one of the plots while a Duraflex membrane was used in the other. The length of the plots was so chosen to accommodate the toe drain as well as the next

diversion berm and drain approximately 12 m above the toe drain.
Since the drainage layer was daylighted into these drains, the
length of the slope in between two diversion berms was considered
critical from the point of view of stability. The width of the test
plot was so chosen as to achieve a length to width ratio of
approximately 2:1, and also to accommodate two widths of the
geomembrane and, hence, a field seam. The slope of the plots was
the design slope of 25 percent. The construction of these plots
demonstrated the feasibility of construction of the cover system as
designed, including the required quality control and quality
assurance efforts. The construction work started in September 1988.
The sideslopes were brought to the required grade using on-site
soils and the lime sludge secondary hydraulic barrier was
constructed at both test plot locations. Two lime sludge layers
were constructed, each 57.9 m (190 ft) long, 15.5 m (51 ft) wide and
5.2 m (17 ft) apart in the center. The placement of geomembrane
(different for the two plots), drainage sand, rooting layer and
topsoil completed the earthwork. The diversion berms and drainage
ditches were constructed as detailed in Figure 2. Two observation
windows, one for each plot, were constructed along the edge of the
membrane in order to observe any slip over the lime sludge. Two
settlement plates were also incorporated in order to monitor
landfill settlements. After the test plots were constructed, the
soils were seeded with winter wheat because it was late for grass
seeding. Erosion control measures, such as erosion mats, were also
used. These plots have been inspected on a monthly basis from
October 1988 through October 1989. Based on the inspections to
date, it has been concluded that the test plots have performed
extremely well. There are no indications of slope movements. A
small settlement of the fill was observed during the month of
September 1989, but this has not caused any problem with the cover
system. No relative movement between the geomembranes and the lime
sludge was observed. In short, the cover system has so far been
proved to be stable. The observations will continue through June
1990. More details of the test plots can be seen in Reference [2].

SUMMARY

    A comprehensive investigation of the stability of a landfill
final cover incorporating a geomembrane was conducted. Certain
parts of the investigation are still continuing. The investigation
was conducted in three phases. In the first phase, the frictional
characteristics between two geomembranes and the drainage sand, as
well as between the cover materials and acrylic, was studied. On
the basis of this study the membrane for Phase II of the
investigation was selected. The second phase consisted of testing a
full scale element of the final cover on a tilt table to investigate
the stability of the cover soils under freeze-thaw conditions as
well as a simulated storm condition. The study was also extended to
determine the factor of safety of the slope under rainfall
conditions and to calibrate two stability analysis methods. Phase
III consisted of constructing two test plots at Anoka Regional
Sanitary Landfill and observing their behavior for over one year.
The conclusions drawn from these investigations are:

1. Both the textured HDPE membrane and the VLDPE membrane
    investigated have better frictional characteristics than the
    reported values of smooth membranes.

2. The test plots constructed using both membranes have performed well for over one year.

3. The tilt table tests show that both infinite slope model and finite element method can be used to estimate the factors of safety for the type of final cover system investigated with reasonable accuracy.

4. The cover system, as tested, has adequate factor of safety against failure. This was determined using stability analysis techniques which were calibrated in this investigation.

5. The freeze-thaw tests showed that the cover system is stable at the designed slopes.

ACKNOWLEDGEMENTS

Richard Lacey and Mahiro Shettima of J&L Testing Company performed the Phase II tests and the stability analysis, respectively. Phase I tests were performed by Dean Ferry of J&L Testing Company. Michael M. Soletski of Foth & Van Dyke and Associates Inc., and personnel of Waste Management of Minnesota, Inc., Waste Management of Wisconsin Inc., and Waste Management, Inc., participated in the construction of the test plots.

REFERENCES

[1]   Williams, N.D. and Houlihan, M.F., "Evaluation of Interface Friction Properties Between Geosynthetics and Soils" Proceedings, Geosynthetics 87, New Orleans, February 1987.

[2]   First Report (1989) and Final Report (1990) on Anoka Regional Sanitary Landfill Soil/Synthetic Final Cover Testing Program, submitted by Foth & Van Dyke to Minnesota Pollution Control Agency.

Samuel B. Levin and Mark D. Hammond

EXAMINATION OF PVC IN A 'TOP CAP' APPLICATION

REFERENCE: Levin, S. B. and Hammond, M. D.,
"Examination of PVC in a 'Top Cap' Application,"
Geosynthetic Testing for Waste Containment
Applications, ASTM STP 1081, Robert M. Koerner,
Ed., American Society for Testing and Materials,
Philadelphia, 1990.

ABSTRACT: The PVC liner installed over Phase I of the Dyer
Boulevard Landfill provides us with an opportunity to examine
the material after 5+ years of exposure to landfill gas and
other environmental stresses in a top cap application. Test
results for samples extracted from the cap are compared to
test results obtained at the time of installation, to material
properties included within the original material specification,
and to material properties from a 'control' sample of excess
PVC material from this closure project kept in a warehouse
since 1983/84.

KEYWORDS: landfill closure, landfill lining, geosynthetics
performance

INTRODUCTION

The Dyer Boulevard Landfill services all of Palm Beach County,
Florida. It is located over the Turnpike Aquifer, an important
local source of drinking water. When initially opened in 1968,
it was one of several unlined landfills operated within the County.
With time, it became a major disposal area for both municipal
solid waste and sewage sludge.

In late 1970's, evidence of contamination of the shallow
aquifer surrounding this high rise landfill was observed within
the site's monitoring network. Under the terms of a consent
agreement between Palm Beach County and the Florida Department
of Environmental Regulation, executed in 1982, it was agreed that:

Mr. Levin is Assistant Manager of the Solid Waste/Resource
Recovery Division of Post, Buckley, Schuh & Jernigan, Inc., 800
North Magnolia Avenue, Suite 600, Orlando, Florida 32803. Mr.
Hammond is Director of Operations, Solid Waste Authority of Palm
Beach County, 5114 Okeechobee Boulevard, Suite 2-C, West Palm
Beach, Florida 33417.

Palm Beach County and the Florida Department of Environmental Regulation, executed in 1982, it was agreed that:

° The Phase I (unlined) landfill was to be closed in an environmentally sound manner.

° A new lined landfill was to be developed to provide for future solid waste disposal capacity.

° An alternative means of wastewater treatment plant sludge and septic tank pumpings disposal was to be utilized.

Post, Buckley, Schuh & Jernigan, Inc. (PBS&J) was selected by Palm Beach County, and subsequently by the Solid Waste Authority of Palm Beach County (SWA) when it assumed responsibility for the landfill in 1983, to develop the closure design for the existing 190 acres ± of landfill cells in Phase I.

The closure design developed for the site included a low permeability 'top cap', well vegetated side slopes, and an integrated drainage system to capture and remove surface runoff, reducing percolation and subsequent leachate generation. The top cap was designed for installation over slopes of less than 10 percent, based on water balance calculations which indicated that only minimal percolation was anticipated through well vegetated landfill side slopes.

The selection of a liner material for use in the top cap at the Dyer Boulevard Landfill proved to be an arduous task, with properties of various materials reported in differing units, or obtained using differing test methods. Suppliers assisted in the selection process by noting the superior properties of some liner materials relative to competing materials. Plasticizer loss, ultraviolet degradation, questionable chemical resistance with respect to landfill gas exposure, and more limited elongation properties were cited as reasons to consider materials other than PVC. Environmental stress cracking, seaming difficulties, and poor strength characteristics upon exposure to bidirectional forces were cited as reasons to consider materials other than High Density Polyethylene (HDPE).

PURPOSE AND SCOPE

During the past several years, geosynthetics testing has matured to the point at which properties of virgin materials are widely available and in many cases readily comparable. Data concerning the properties of liner materials which have been in service remain scarce, although there is a growing body of information concerning exposure in the laboratory to simulated in-service environments [1], [2].

The top cap material in-service at Dyer Boulevard provides an opportunity to examine the properties of material which has been in service for over five years. Properties of this material will be compared to the properties determined by quality assurance

testing during its manufacture and installation, in an attempt
to assess the change in properties resulting from material exposure.
Excess material stored in a warehouse since its purchase in 1983/84
will serve as a control.

DESIGN AND SPECIFICATIONS

The site closure design included the placement of PVC sheet,
soil bedding and cover material, a passive landfill gas venting
system, drainage improvements, and seeding, mulching and sodding
of the completed landfill. A typical cross section through the
final cover is provided in Figure 1. Physical properties specified
for the 20 mil PVC liner are presented in Table 1. Tensile strength
at the seam was required to be at least 80 percent of that of
the parent material, or 1760 psi. Also required in the
specification was the sampling and testing of the production run
for tensile strength and elongation at break.

General Contractor Crabtree Construction Company, Inc.,
purchased 3.32 million square feet of PVC material, Product Number
1951, manufactured by Dynamit Nobel of America, Inc. and fabricated
by the Watersaver Company, Inc. The surface of the subgrade
prepared as liner bedding was treated with Hyvar X-L Herbicide
prior to placement and seaming of the liner panels. The coarse
grained sand used for liner bedding and backfill was obtained
by dredge from a near site borrow area.

One hundred and sixty-six panels of PVC, most of which measured
400 feet by 70 feet, were installed by Wright/Kohli Construction
Company, a specialty liner subcontractor. Watersaver WS-70 splicing
solvent was used for seaming the panels together.

TABLE 1 -- Specified Physical Properties for 20 Mil PVC [3]

Physical Property	Value	ASTM Test
Thickness	± 10%	D 1593
Specific Gravity	1.24 - 1.30	D 792
Tensile Strength	2200 psi	D 882 or D 412
Elongation	300%	D 882 or D 412
100% Modulus	1000 psi	D 882
Graves Tear	270 lbs/in	D 1004
Water Extraction	0.35% (max)	D 1239
Volatility	0.70% (max)	D 1203
Impact Cold Crack	-20°F	D 1790
Dimensional Stability	5% max @ 212°F	D 1204-54

The liner material was placed and covered from January through
October, 1984.

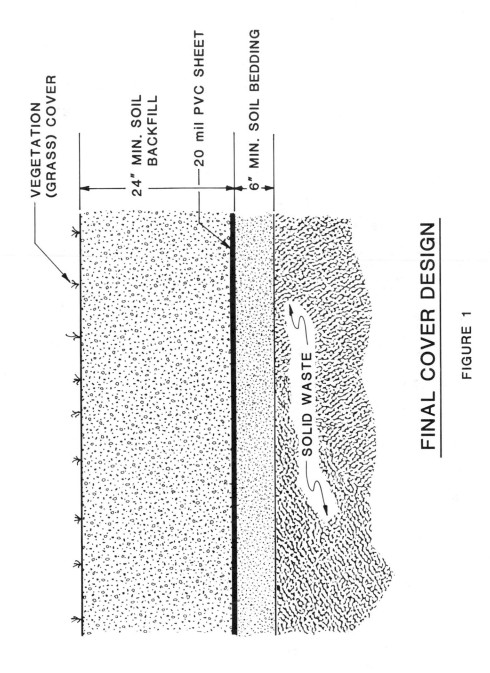

VEGETATION (GRASS) COVER

24" MIN. SOIL BACKFILL

20 mil PVC SHEET

6" MIN. SOIL BEDDING

SOLID WASTE

FINAL COVER DESIGN

FIGURE 1

PVC SHEET (LINER) TESTING

Testing of the properties of the PVC sheet installed over the Dyer Boulevard landfill included:

° Quality control (QC) testing of the parent material at the time of manufacture by Dynamit Nobel of America, Inc.

° Non destructive (air lance) testing of all field seams during the time of installation.

° Destructive testing of the parent material, factory and field seam samples obtained throughout the ten month installation period.

Table 2 provides a summary of the manufacturer's QC data. Please note that the data presented is the average for a number of rolls tested. One sample was obtained for testing purposes for each 86,000 square feet of manufactured product.

Destructive testing of field samples taken throughout the liner installation process was performed by QC Metallurgical Inc., Hollywood, Florida. A total of 236 tensile strength tests were performed on field samples. These test results are summarized in Table 3. All tensile tests were performed in shear.

LINER RESAMPLING/TESTING PROGRAM

In December, 1989, the soil overburden was carefully removed from the top cap in several locations of the Phase I area to expose the liner material. According to the Solid Waste Authority's manager of landfills, the exposed material appeared to exhibit few signs of degradation upon visual inspection.

The surface of the material exhibited minor undulations, a few millimeters in size, where the material apparently elongated to conform to the surfaces of soil particles. The samples were highly pliable, exhibiting no apparent brittleness.

Eight coupons of the parent material and one field seam coupon were sliced from the cap by the Solid Waste Authority's manager of landfills. Added to the package of nine samples was a coupon sliced from a factory panel of PVC stored by the Solid Waste Authority as surplus in a warehouse since its purchase in 1983/84 for capping the Phase I Dyer Boulevard site. The ten samples were transmitted to Geosyntec, Inc., Boynton Beach, Florida for laboratory testing.

A summary of test results for these samples is provided in Table 4. The tests were selected to correspond with tests performed by the manufacturer in 1984. Low temperature impact testing was not performed due to its limited applicability to the sub-tropical West Palm Beach environment. Dimensional stability was not determined due to an insufficient quantity of sample material.

TABLE 2 -- Manufacturer's QC Data[a]

Properties	ASTM Test Method	Manufacturer's Specification	Test Result (Ave.11 Rolls)	Test Result (Ave.6 Rolls)	Test Result (Ave.13 Rolls[b])	Test Result (Ave.7 Rolls)
Thickness (mils)	D-1593	20 ± 5%	19.9	19.5	19.8	19.9
Specific Gravity (min.)	D-792	1.23	1.25	1.25	1.25	1.25
Tensile Strength (psi)	D-882	2400	MD 2999 TD 2768	MD 2836 TD 2692	MD 2487 TD 2464	MD 2977 TD 2781
Modulus @ 100% Elong. (psi min.)	D-882	1000	MD 1446 TD 1333	MD 1270 TD 1202	MD 1384 TD 1332	MD 1291 TD 1203
Elongation, % min.	D-882	300	MD 414 TD 426	MD 423 TD 435	MD 336 TD 354	MD 453 TD 449
Tear Strength (lbs.)	D-1004	5.5	MD 7.46 TD 7.92	Not Reported	8.51 9.18	Not Reported
Low Temperature Impact °F	D-1790	-20	Pass	Pass	Pass	Pass
Volatile Loss, max. (@70°C for 24 hrs.)	D-1203	1.0	0.79	0.70%	0.87%	1.13%
Water extraction, % Loss max (104°F, for 24 hrs.)	D-1239	0.3	0.08	0.18	0.15	0.30
Dimensional Stability % change, max. 212°F for 15 min.	D-1204	±5.0	MD -1.8 TD 0.6	MD -1.8 TD 1.7	MD -1.8 TD 0.86	MD -1.27 TD 0.51

a Provided by the Watersaver Company, Inc.
b Number of rolls tested estimated based upon reported material weight.

MD - Machine Direction
TD - Transverse Direction

TABLE 3 -- Summary of Tensile Test Data for Field Samples of PVC when Installed

Sample Type	Sample Quantity	Specified Values (psi)	Range of Results (psi)	Average Value (psi)	Sample Standard Deviation
Field Seam in Shear	195	1760	1490 - 2550	1937	165
Factory Seam in Shear	30	1760	1825 - 2568	2132	186
Parent Material	11	2200	2040 - 4750	2857	712

TABLE 4 -- Liner Resampling/Testing Program

	Specific Gravity	Tensile Strength (psi)	Modulus @ 100% Elongation (psi)	Elongation (%)	Tear Strength (lbs.)	Volatile Loss (%)	Water Extraction (%)
ASTM Test Method	D-792	D-882	D-882	D-882	D-1004	D-1203	D-1239
Field Seam	N/A	2204 shear 1140 peel	N/A	407	N/A	N/A	N/A
Field Panel 1	1.265	MD 2744 TD 2642	MD 1865 TD 1721	MD 341 TD 381	MD 10.1 TD 9.3	1.33	2.81
Field Panel 2	1.282	MD 3072 TD 2642	MD 1865 TD 1721	MD 341 TD 381	MD 10.1 TD 9.3	0.40	1.08
Field Panel 3	1.267	MD 2740 TD 2544	MD 1968 TD 1828	MD 341 TD 334	MD 9.9 TD 9.6	1.41	2.06
Field Panel 4	1.264	MD 2614 TD 2633	MD 2043 TD 1904	MD 340 TD 324	MD 9.5 TD 9.8	2.02	1.85
Field Panel 5	1.267	MD 2634 TD 2327	MD 1884 TD 1879	MD 311 TD 309	MD 9.5 TD 8.6	3.26	1.99
Field Panel 6	1.252	MD 2498 TD 2360	MD 1667 TD 1794	MD 356 TD 270	MD 8.8 TD 8.9	4.23	2.63
Field Panel 7	1.270	MD 2446 TD 2149	MD 1746 TD 1672	MD 309 TD 331	MD 9.2 TD 8.5	0.85	1.69
Field Panel 8	1.278	MD 2627 TD 2628	MD 1763 TD 1781	MD 362 TD 345	MD 8.9 TD 8.6	0.96	2.58
Warehouse Panel	1.279	MD 2662 TD 2439	MD 1557 TD 1476	MD 429.4 TD 417.4	MD 8.3 TD 7.8	0.78	1.29

MD - Machine Direction
TD - Transverse Direction

TABLE 5 -- Summary of Test Results

		Parent Material (Manufacturer's QC Data)[a]	Parent Material (from Warehouse)	Parent Material (field samples December 1989)
Specific Gravity	Range	1.25 - 1.25	1.28	1.25 - 1.28
	Ave.	1.25		1.27
Tensile Strength (psi)	MD Range	2487 - 2999	2662	2498 - 3072
	MD Ave.	2789		2672
	TD Range	2464 - 2781	2439	2149 - 2729
	TD Ave.	2651		2502
Break Elongation (%)	MD Range	336 - 453	429	309 - 362
	MD Ave.	395		335
	TD Range	354 - 449	417	270 - 381
	TD Ave.	407		327
Tear Resistance (lbs.)	MD Range	7.5 - 8.5	8.3	8.9 - 10.1
	MD Ave.	8.0		9.5
	TD Range	7.9 - 9.2	7.8	8.5 - 9.8
	TD Ave.	8.6		9.1
Secant Modulus (psi)	MD Range	1270 - 1446	1557	1667 - 2243
	MD Ave.	1366		1897
	TD Range	1202 - 1333	1476	1672 - 2027
	TD Ave.	1287		1826
Volatile Loss (%)	Range	0.70 - 1.13	0.78	0.4 - 4.23
	Ave.	0.87		1.81
Water Extraction (%)	Range	0.08 - 0.30	1.29	1.69 - 2.81
	Ave.	0.16		2.09

[a] Average values presented here are average of the average QC data reported by the manufacturer, weighed by the number of rolls tested.

COMPARISON/EVALUATION OF TEST RESULTS

An insufficient number of field seam samples (one) and warehouse parent material samples (one) were retested to provide for a reliable comparison of test results. Data from the testing of these samples is provided for information only. Emphasis will be placed on comparing the data from the factory QC parent material test results with testing of the eight parent material samples obtained in December, 1989. This data is presented for comparison in Table 5.

The specific gravity of the 'aged' PVC appears to be somewhat greater than that at the time of manufacture. This increase is consistent with a loss of plasticizer in the aged samples, since the specific gravity of the PVC resin (approximately 1.4) exceeds that of the plasticizer (approximately 0.98). Based on the apparent increase in specific gravity, the aged samples have lost approximately 13 percent of their initial (factory) plasticizer content.

Plasticizer loss would expectedly be evidenced by an increase in tensile strength. The aged samples did not reflect this, with a loss in the average tensile strength of about 4 and 6 percent, machine direction and transverse direction respectively.

Conversely, break elongation results followed a highly predictable path. The average aged sample elongation was 15 percent (machine direction) and 20 percent (transverse direction) less than that reported on the factory QC sheets.

The tear resistance test, as its name implies, is a measure of the force necessary to initiate a tear in the plastic sheet. The averaged test sample tear resistance values for the aged samples exceeded the average values observed at the time of manufacture by 6 and 19 percent.

Secant modulus at 100% strain provides a measure of the stiffness of the sheet, and should increase as plasticizer is lost from the material. The modulus at 100% strain of the aged samples averaged 42 and 39 percent above the average modulus data reported at the time of manufacture.

The volatile loss test is subject to some variability, depending on the type of plasticizer used [4],[5]. The large difference between the average volatile loss tested at the time of manufacture and that of the aged samples likely exceeds variations inherent in the test. The average volatile loss of the aged samples exceeded the average loss reported by DNA by over 100 percent.

An even larger increase was observed between average values for the water extraction test. The average loss values for the aged samples exceeded those reported at the time of manufacture by over 1000 percent.

DISCUSSION

The above described apparent changes in properties of the PVC sheet, with the possible exception of tensile strength, are consistent with changes associated with plasticizer loss. Since designers, owners, and operators of solid waste landfills are most concerned with the long term effectiveness of the top cap, what we really wish to know is of what point will plasticizer loss be sufficient to result in cap failure, and how long will it take for this failure to occur?

It would appear from both visual observation and laboratory testing that for a South Florida landfill, this period exceeds five years. With few exceptions, the properties which relate to the ability of the PVC to continue to function as a low permeability moisture barrier, (tensile strength, elongation, tear resistance, secant modulus) exceeded the originally specified values in each of the eight aged samples.

How long beyond this five year period will the PVC sheet continue to provide its intended top cap function? Extrapolation of the data obtained by testing the eight aged samples is tenuous, at best.

Some insight may be provided through review of the comprehensive study performed by Morrison and Starbuck [5]. Testing was performed on 10 mil PVC linings within eight canals which had been in service for periods ranging from 0.6 to 19 years. The changes in properties observed by Morrison and Starbuck were mostly consistent with those observed in the eight landfill top cap samples, and were attributed to plasticizer loss. The landfill top cap environment does not appear to have been more hostile to PVC than the canal lining environment, although the difference in thickness (10 mil for canal linings, 20 mil for the top cap) may have impacted the relative magnitude of the observed/reported changes in properties.

The Morrison and Starbuck study provides data for 10 mil canal lining samples after as many as 18 and 19 years in service. Loss in elongation for the 19 year old sample is as high as 63.3%, with a corresponding loss of plasticizer of 45.7%. These property changes were not sufficient to cause failure of the canal lining.

An additional potential difference between the landfill top cap environment and the canal lining environment is temperature. The top cap may be exposed to elevated landfill gas temperatures resulting from waste decomposition. Elevated temperatures are a driving force for plasticizer loss. In-situ temperatures were not noted in the canal lining study, making comparisons with the top cap environment more difficult.

How long a service life is required for a landfill top cap? In Florida, the post closure period for a solid waste landfill extends a minimum of 20 years [6]. The latest proposed federal rules [7] require a post closure period which extends for a minimum of 30 years.

The stresses imposed on a top cap in a solid waste landfill, subsequent to its installation, are in our opinion primarily a result of differential waste settlement. Sufficient elongation to accommodate this settlement is of paramount importance.

The rate and uniformity at which landfills settle is highly variable depending to a large extent on the type of waste accepted, compaction procedures, moisture content, and overall landfill height. Landfill settlement tends to be most pronounced in the inital several years after closure. It is during this period that the PVC exhibits its greatest elongation properties.

SUMMARY AND CONCLUSIONS

The PVC top cap at the Dyer Boulevard Landfill appears to have maintained its integrity after its initial 5+ years in service. Test values from sample coupons extracted from the top cap differ from test values for sample coupons taken from the parent material at the time of manufacture. These apparent property changes are consistent with those attributed to plasticizer loss, published for PVC which has been in service as canal lining, although the canal lining and top cap environments may differ.

The aging of PVC in both the top cap and canal environments appears to be consistent with changes in properties associated with plasticizer loss although other factors in the landfill environment or canal environment may be responsible for these observed changes. In the canal environment, the PVC exhibited significant loss of plasticizer over 18 and 19 years of exposure. Corresponding elongation properties were reduced by as much as 63%. The material was still serving its intended function despite these changes in properties. Additional long term data (preferably from landfill top cap environments) will assist in projecting the anticipated service life of PVC for landfill closures.

ACKNOWLEDGEMENTS

We wish to express our sincere appreciation to the Solid Waste Authority of Palm Beach County for funding the liner testing program. Additional thanks are due to Art Arena, Technical Director, Huls America, Inc., and Richard Dickenson, a technical consultant to Huls America, Inc.

REFERENCES

[1] W.R. Morrison and L.D. Parkhill, Evaluation of Flexible Membrane Liner Seams after Chemical Exposure and Simulated Weathering, U.S. Bureau of Reclamation, Engineering and Research Center, Denver, Colorado, Feburary 1987.
[2] Haxo, H.E., et.al., Liner Materials Exposed to Municipal Solid Waste Leachate, Third Interim Report, EPA-600/2-79-038, July 1979.

[3] Post, Buckley, Schuh & Jernigan, Inc., Contract Specifications
     for Dyer Boulevard Landfill Closure for Palm Beach County,
     Florida, June 1983.
[4] ASTM, 1989 Annual Book of ASTM Standards, Section 8, Volume
     08.01.
[5] W.R. Morrison and J.G. Starbuck, Performance of Plastic Canal
     Linings, U.S. Bureau of Reclamation, Engineering and Research
     Center, Denver, Colorado, January 1984.
[6] Florida Administration Code, Chapter 17-701.07.
[7] Federal Register, Volume 53, No. 168, Tuesday, August 30,
     1988/Proposed Rules, Page 33408.

Author Index

Subject Index